S0-AGC-925

Praise for
THE TELOMERE EFFECT

"Elizabeth Blackburn and Elissa Epel have discovered that telomeres, the capping structure at the end of your DNA that make up your chromosomes, do not simply carry out the commands issued by your genetic code. Telomeres are listening to you. They absorb the instructions you give them. They respond to you being stressed and to your being relaxed, to your being sad or your being happy. Thus, telomeres contribute to the state of your brain, to your mood, to the speed of your aging, and to your risk for neurodegenerative diseases. In other words, we can change the way we age at the most elemental cellular level. So if you want to keep your brain sharp, you have to know about your telomeres and stay in touch with them. This book shows you how you can do this, and it does so in a way that is both intellectually exhilarating yet easily accessible to all. This book will become a classic. It is one of the most exciting books in biology to emerge in the last decade."

—Eric Kandel, Nobel laureate, author of *In Search of Memory: The Emergence of a New Science of Mind*

"Improving public health requires that people know the truth about their own lives. Blackburn and Epel reveal the discovery of how cells age and how certain forces in our lives cause us to get sick and age prematurely. *The Telomere Effect* explains the often-invisible things that affect all of our lives, giving us a fresh new level of awareness and helping us make better choices individually and socially for greater health and longevity. In short, it will change the way we think of aging and disease."

—David Kessler, MD, former FDA commissioner, author of the *New York Times* bestseller *The End of Overeating* and *Capture*

"Using both science and personal stories, Blackburn and Epel demonstrate that how we live each day has a profound effect not just on our health and well-being but how we age, as well. It's a manual for how to live younger and longer. Spoiler alert: sleep is a key element. *The Telomere Effect* is a book that will help you thrive at every level."

—Arianna Huffington

"Elizabeth Blackburn and Elissa Epel have discovered a revolutionary set of findings that can transform the way we live our lives, shaping the very health of our cells by how we use our minds. These pioneers of well-being unveil a story of the power of our interpersonal connections—in romance, friendship, and child-parent relationships—to slow the rate of cell aging.

"These powerful discoveries are made useful in your day-to-day life with a wealth of science-based suggestions, which will delight your mind, enrich your day, and improve your health."

—Daniel J. Siegel, MD,
author of *Mindsight* and *Brainstorm*

"Blackburn and Epel lay out a road map for thriving as we age by eloquently illuminating the intricate relationships between the psychology and the biology of aging. Drawing on telomere science, they empower and inspire readers to enhance their healthspans. The authors point to realistic possibilities long life affords, in language that is accessible, informative, and highly engaging."

—Laura L. Carstensen, PhD, professor of psychology,
founding director of the Center on Longevity at
Stanford University, author of
A Long Bright Future

"The Blackburn-Epel 'dream team' has condensed a massive body of complex scientific data into a highly readable, nontechnical 'how-to' manual on strategies that will help anyone who is human: a truly extraordinary gift to all of us who want to enhance our health, no matter at what stage of life we are in."

—Rita B. Effros, PhD, professor, David Geffen School of Medicine at UCLA, 2015 president, Gerontological Society of America

"The authors of this fascinating book show how the telomeres in our bloodstreams are responsive to many aspects of our daily existence. In these pages, telomeres become the nexus of an important discussion of vulnerability and resilience to influences of our social and physical environment and the important role of the mind-body connection. In our future, it may be that telomere monitoring will help us pursue better health—a new frontier waiting to be explored. Regardless, you will learn a lot that can benefit your 'healthspan.'"

—Bruce McEwen, PhD, professor of neuroscience, the Rockefeller University, author of *The End of Stress as We Know It*

"Dr. Elizabeth Blackburn is *the* expert on telomeres, which are the tips that protect our chromosomes and correlate remarkably with health and longevity. Her and Dr. Epel's scientific discoveries and their potential importance for our health, both individually and collectively, are profound, and their apparent relationship to stress opens up an exciting array of potential healthy lifestyle changes."

—Lee Goldman, MD, chief executive of Columbia University Medical Center, author of *Too Much of a Good Thing: How Four Key Survival Traits Are Now Killing Us*

"The breakthrough research that Drs. Elizabeth Blackburn and Elissa Epel have conducted has created a dramatic shift in our understanding of what is possible in terms of human health and longevity. Telomeres, the ends of your DNA, affect how quickly your cells age and die. As your telomeres get shorter, your life gets clouded with disease.

"Drs. Blackburn and Epel are the central researchers in the discovery of telomeres, their profound effect on health, and the myriad ways in which lifestyle choices can improve cellular aging. They have collaborated with researchers worldwide on studies from understanding cell-aging machinery to chemical exposures to mental-training classes aimed to improve cellular health. A study we collaborated on showed, for the first time, that comprehensive lifestyle changes may actually increase the length of our telomeres, thus beginning to reverse aging on a cellular level. This book is revolutionary, transforming the way our world thinks about health and living well, disease, and death. This work reveals a stunning picture of healthy aging—it's not simply about individuals; it's about how we are connected to each other, today and through future generations. It is hard to overstate its importance."

Dean Ornish, MD, founder and president, Preventive Medicine Research Institute, clinical professor of medicine, UCSF, *New York Times* bestselling author of *The Spectrum*

"Some of us emphasize social determinants of health; others of us emphasize behaviors such as diet and exercise; still others look at psychology and health. What if we had a coherent and readily comprehensible way of understanding the biology that links all of these to health and disease and to length and quality of life? We would not only gain better understanding of causes of health and disease, but what to do to improve things. As Blackburn and Epel lay out beautifully and clearly in this wonderful book, telomere length provides such a unifying biological mechanism. The authors take cutting-edge science and make it fascinating and intelligible for interested reader and expert alike. More than that, we warm to their humanity."

—Professor Sir Michael Marmot,
president of the World Medical Association, director,
University College London Institute of Health Equity, author
of *The Health Gap: The Challenge of an Unequal World*

"At last we are closing in on the intertwined biological, behavioral, and social influences on why some people thrive in good health while others are more likely to stumble and fall. Always educational and sometimes poetic, *The Telomere Effect* brings us a fascinating analysis from two of the world's best researchers on behavior, health, and longevity.

"Avoiding quick-fix exhortations like New Year's resolutions destined to be melting by spring, Blackburn and Epel explain the longer-term life patterns that play a role in longer telomeres, longer periods of good health, and longer life.

"This excellent book avoids the trap of seeing all stress and challenge as bad, and instead provides a nuanced understanding that trials and tribulations are not inevitably a health threat, because challenge can build resilience. The study of telomeres helps us understand what protects and toughens our cells. This is cutting-edge longevity science."

—Howard S. Friedman, PhD, distinguished professor
at the University of California, Riverside, author of
*The Longevity Project: Surprising Discoveries for Health and
Long Life from the Landmark Eight-Decade Study*

"*The Telomere Effect* gives us, in high relief and with exactly the practical level of detail we need, the long and the short of a new science revealing that how we live our lives, both inwardly and outwardly, individually and collectively, impinges significantly on our health, our well-being, and even our longevity. Mindfulness is a key ingredient, and importantly, issues of poverty and social justice are shown to clearly come into play as well. This book is an invaluable, rigorously authentic, and at its core, exceedingly compassionate and wise contribution to our understanding of health and well-being."

—Jon Kabat-Zinn, author of *Full Catastrophe Living*
and *Coming to Our Senses*

THE
TELOMERE
EFFECT

THE TELOMERE EFFECT

A REVOLUTIONARY APPROACH TO
LIVING YOUNGER, HEALTHIER, LONGER

Elizabeth Blackburn, PhD
Elissa Epel, PhD

GRAND CENTRAL
PUBLISHING

NEW YORK   BOSTON

This book is designed to help you understand the new science of telomeres and to help you make informed lifestyle choices; it is not meant to replace formal medical treatment by a physician or other licensed health care provider. You should regularly consult a physician in matters relating to your health and particularly with respect to any symptoms that may require diagnosis or medical attention.

Copyright © 2017 by Elizabeth Blackburn and Elissa Epel

Cover design by Jeff Miller, Faceout Studio
Cover copyright © 2017 by Hachette Book Group, Inc.

Hachette Book Group supports the right to free expression and the value of copyright. The purpose of copyright is to encourage writers and artists to produce the creative works that enrich our culture.

The scanning, uploading, and distribution of this book without permission is a theft of the authors' intellectual property. If you would like permission to use material from the book (other than for review purposes), please contact permissions@hbgusa.com. Thank you for your support of the authors' rights.

Grand Central Publishing
Hachette Book Group
1290 Avenue of the Americas, New York, NY 10104
grandcentralpublishing.com
twitter.com/grandcentralpub

First Edition: January 2017

Grand Central Publishing is a division of Hachette Book Group, Inc. The Grand Central Publishing name and logo is a trademark of Hachette Book Group, Inc.

The publisher is not responsible for websites (or their content) that are not owned by the publisher.

The Hachette Speakers Bureau provides a wide range of authors for speaking events. To find out more, go to www.hachettespeakersbureau.com or call (866) 376-6591.

Illustrations by Colleen Patterson of Colleen Patterson Design

Library of Congress Cataloging-in-Publication Data

Names: Blackburn, Elizabeth H. (Elizabeth Helen), 1948– author. | Epel,
   Elissa S. (Elissa Sarah), 1968– author.

Title: The telomere effect : a revolutionary approach to living younger, healthier, longer / by Elizabeth
   Blackburn, Elissa Epel.

Description: New York : Grand Central Publishing, [2017] | Includes
   bibliographical references and index.

Identifiers: LCCN 2016028884| ISBN 9781455587971 (hardcover) |
   ISBN 9781455541713 (hardcover large-print) | ISBN 9781478940425 (audio cd) |
   ISBN 9781478940432 (audio download) | ISBN 9781455587964 (ebook)

Subjects: | MESH: Telomere—physiology | Aging—genetics

Classification: LCC QH600.3 | NLM QU 470 | DDC 572.8/7—dc23 LC record available at
https://lccn.loc.gov/2016028884

ISBNs: 978-1-4555-8797-1 (hardcover), 978-1-4555-4171-3 (large print),
978-1-4555-8796-4 (ebook)

Printed in the United States of America

LSC-H

10 9 8 7 6 5 4 3 2 1

I dedicate this book to John and Ben,
the lights of my life, who simply make everything
for me worthwhile. —EHB

I dedicate this book to my parents, David and Lois,
who are an inspiration in how they live fully
and vibrantly, in their almost ninth decade of life,
and to Jack and Danny, who make my cells happy. —ESE

Contents

PART III
HELP YOUR BODY PROTECT ITS CELLS

PART IV
OUTSIDE IN: THE SOCIAL WORLD SHAPES
YOUR TELOMERES

Authors' Note: Why We Wrote This Book

With a life span of 122 years, Jeanne Calment was one of the longest-living women on record. When she was eighty-five, she took up the sport of fencing. She was still riding a bike into her triple digits.[1] When she turned one hundred, she walked around her hometown of Arles, France, thanking the people who'd wished her a happy birthday.[2] Calment's relish for life captures what we all want: a life that is healthy right up to the very end. Aging and death are immutable facts of life, but how we live until our last day is not. This is up to us. We can live better and more fully now and in our later years.

The relatively new field of telomere science has profound implications that can help us reach this goal. Its application can help reduce chronic disease and improve wellbeing, all the way down to our cells and all the way through our lives. We've written this book to put this important information into your hands.

Here you will find a new way of thinking about human aging. One current, predominant, scientific view of human aging is that the DNA of our cells becomes progressively damaged, causing cells to become irreversibly aged and dysfunctional. But which DNA is damaged? Why did it become damaged? The full answers aren't known yet, but the clues are now pointing strongly toward telomeres as a major culprit. Diseases can seem distinct because they involve very different organs and parts of the body. But new scientific and clinical findings have crystallized into a new concept. Telomeres throughout the body shorten as we age, and this underlying mechanism contributes to most diseases of aging. Telomeres explain how we run out of the abilty to replenish tissue (called replicative senescence). There are other ways cells become dysfunctional or die early, and there are other factors that contribute to human aging. But telomere attrition is a clear and an early contributor to the aging process, and—more exciting—it is possible to slow or even reverse that attrition.

We've put the lessons from telomere research into the full story, as it is unfolding today, in language for the general reader. Previously this knowledge has been available only in scientific journal articles, scattered in bits and pieces. Simplifying this body of science for the public has been a great challenge and responsibility. We could not describe every theory or pathway of aging or lay out each topic in fine scientific detail. Nor could we state every qualification and disclaimer. Those issues are detailed in the scientific journals where the original studies were published, and we encourage interested readers to explore this fascinating body of work, much of it cited in this book. We have also written a review article covering the latest research on telomere biology, published in the peer-reviewed scientific journal *Science*, which will give you several good directions into the molecular-level mechanisms.[3]

Science is a team sport. We have been truly privileged to participate in research with a broad range of scientific collaborators from different disciplines. We have also learned from research teams from all over the world. Human aging is a puzzle made up of many pieces. Over several decades, new pieces of information have each added a critical part to the whole. The understanding of telomeres has helped us see how the pieces fit together—how aged cells can cause the vast array of diseases of aging. Finally a picture has emerged that is so compelling and helpful that we felt it was important to share it broadly. We now have a comprehensive understanding of human telomere maintenance, from cell to society, and what it can mean in human lives and communities. We are sharing with you the basic biology of telomeres, how they relate to disease, to health, to how we think, and even to our families and communities. Putting together the pieces, illuminated by knowledge of what affects telomeres, has led us to a more interconnected view of the world, as we share with you in the last section of the book.

Another reason we've written this book is to help you avoid potential risks. The interest in telomeres and aging is growing exponentially, and while there is some good information in the public domain, some of it is misleading. For example, there are claims that certain creams and supplements may elongate your telomeres and increase your longevity. These treatments, if they actually work in the body, could potentially increase your risk of cancer or have other dangerous effects. We simply need larger and longer studies to assess these potential serious risks. There are other known ways to improve your cell longevity, without risk, and we have tried to include the best of them here. You won't find any instant cures on these pages, but you *will* find the specific, research-supported ideas that could make the rest of your life healthy, long, and fulfilling. While some ideas may not be totally new to you, gaining a deep understanding of the behind-the-scenes reasons for them may change how you view and live your days.

Finally, we want you to know that neither of us has any financial interest in companies that sell telomere-related products or that offer telomere testing. Our wish is to synthesize the best of our understanding—as it stands today—and make it available to anyone who may find it useful. These studies represent a true breakthrough in our understanding of aging and living younger, and we want to thank all who have contributed to the research that we are able to present here.

With the exception of the "teaching story" that appears on the first page of the introduction, the stories in this book are drawn from real-life people and experiences. We are deeply grateful to the people who shared their stories with us. To protect their privacy, we have changed some names and identifying details.

We hope this book is helpful to you, your families, and all who can benefit from these fascinating discoveries.

A Tale of Two Telomeres

It is a chilly Saturday morning in San Francisco. Two women sit at an outdoor café, sipping hot coffee. For these two friends, this is their time away from home, family, work, and to-do lists that never seem to get any shorter.

Kara is talking about how tired she is. How tired she *always* is. It doesn't help that she catches every cold that goes around the office, or that those colds inevitably turn into miserable sinus infections. Or that her ex-husband keeps "forgetting" when it's his turn to pick up the children. Or that her bad-tempered boss at the investment firm scolds her—right in front of her staff. And sometimes, as she lies down in bed at night, Kara's heart gallops out of control. The sensation lasts for just a few seconds, but Kara stays awake long after it passes, worrying. *Maybe it's just the stress*, she tells herself. *I'm too young to have a heart problem. Aren't I?*

"It's not fair," she sighs to Lisa. "We're the same age, but I look older."

She's right. In the morning light, Kara looks haggard. When she reaches for her coffee cup, she moves gingerly, as if her neck and shoulders hurt.

1

But Lisa looks vibrant. Her eyes and skin are bright; this is a woman with more than enough energy for the day's activities. She feels good, too. Actually, Lisa doesn't think very much about her age, except to be thankful that she's wiser about life than she used to be.

Looking at Kara and Lisa side by side, you would think that Lisa really *is* younger than her friend. If you could peer under their skin, you'd see that in some ways, this gap is even wider than it seems. Chronologically, the two women are the same age. Biologically, Kara is decades older.

Does Lisa have a secret—expensive facial creams? Laser treatments at the dermatologist's office? Good genes? A life that has been free of the difficulties her friend seems to face year after year?

Not even close. Lisa has more than enough stresses of her own. She lost her husband two years ago in a car accident; now, like Kara, she is a single mother. Money is tight, and the tech start-up company she works for always seems to be one quarterly report away from running out of capital.

What's going on? Why are these two women aging in such different ways?

The answer is simple, and it has to do with the activity inside each woman's cells. Kara's cells are prematurely aging. She looks older than she is, and she is on a headlong path toward age-related diseases and disorders. Lisa's cells are renewing themselves. She is living younger.

WHY DO PEOPLE AGE DIFFERENTLY?

Why do people age at different rates? Why are some people whip smart and energetic into old age, while other people, much younger, are sick, exhausted, and foggy? You can think of the difference visually:

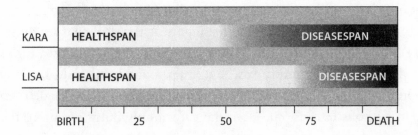

Figure 1: Healthspan versus Diseasespan. Our healthspan is the number of years of our healthy life. Our diseasespan is the years we live with noticeable disease that interferes with our quality of living. Lisa and Kara may both live to one hundred, but each has a dramatically different quality of life in the second half of her life.

Look at the first white bar in figure 1. It shows Kara's healthspan, the time of her life when she's healthy and free of disease. But in her early fifties, the white goes gray, and at seventy, black. She enters a different phase: the diseasespan.

These are years marked by the diseases of aging: cardiovascular disease, arthritis, a weakened immune system, diabetes, cancer, lung disease, and more. Skin and hair become older looking, too. Worse, it's not as if you get just one disease of aging and then stop there. In a phenomenon with the gloomy name *multi-morbidity*, these diseases tend to come in clusters. So Kara doesn't just have a run-down immune system; she also has joint pain and early signs of heart disease. For some people, the diseases of aging hasten the end of life. For others, life goes on, but it's a life with less spark, less zip. The years are increasingly marred by sickness, fatigue, and discomfort.

At fifty, Kara should be brimming with good health. But the graph shows that at this young age, she is creeping into the diseasespan. Kara might put it more bluntly: she is getting old.

Lisa is another story.

At age fifty, Lisa is still enjoying excellent health. She gets older as the years pass, but she luxuriates in the healthspan for a nice, long

3

time. It isn't until she's well into her eighties—roughly the age that gerontologists call "old old"—that it gets significantly harder for her to keep up with life as she's always known it. Lisa has a diseasespan, but it's compressed into just a few years toward the end of a long, productive life. Lisa and Kara aren't real people—we've made them up to demonstrate a point—but their stories highlight questions that are genuine.

How can one person bask in the sunshine of good health, while the other suffers in the shadow of the diseasespan? Can you choose which experience happens to *you*?

The terms *healthspan* and *diseasespan* are new, but the basic question is not. *Why do people age differently?* People have been asking this question for millennia, probably since we were first able to count the years and compare ourselves to our neighbors.

At one extreme, some people feel that the aging process is determined by nature. It's out of our hands. The ancient Greeks expressed this idea through the myth of the Fates, three old women who hovered around babies in the days after birth. The first Fate spun a thread; the second Fate measured out a length of that thread; and the third Fate snipped it. Your life would be as long as the thread. As the Fates did their work, *your* fate was sealed.

It's an idea that lives on today, although with more scientific authority. In the latest version of the "nature" argument, your health is mostly controlled by your genes. There may not be Fates hovering around the cradle, but the genetic code determines your risk for heart disease, cancer, and general longevity before you're even born.

Perhaps without even realizing it, some people have come to believe that nature is *all* that determines aging. If they were pressed to explain why Kara is aging so much faster than her friend, here are some things they might say:

"Her parents probably have heart problems and bad joints, too."

"It's all in her DNA."

"She has unlucky genes."

4

The "genes are our destiny" belief is, of course, not the only position. Many have noticed that the quality of our health is shaped by the way we live. We think of this as a modern view, but it's been around for a long, long time. An ancient Chinese legend tells of a raven-haired warlord who had to make a dangerous trip over the border of his homeland. Terrified that he would be captured at the border and killed, the warlord was so anxious that he woke up one morning to discover that his beautiful dark hair had turned white. He'd aged early, and he'd aged overnight. As many as 2,500 years ago, this culture recognized that early aging can be triggered by influences like stress. (The story ends happily: No one recognized the warlord with his newly whitened hair, and he traveled across the border undetected. Getting older has its advantages.)

Today there are plenty of people who feel that nurture is more important than nature—that it's not what you're born with, it's your health habits that really count. Here's what these folks might say about Kara's early aging:

"She's eating too many carbs."

"As we age, each of us gets the face we deserve."

"She needs to exercise more."

"She probably has some deep, unresolved psychological issues."

Take a look again at the ways the two sides explain Kara's accelerated aging. The nature proponents sound fatalistic. For good or for bad, we're born with our futures already encoded into our chromosomes. The nurture side is more hopeful in its belief that premature aging can be avoided. But advocates of the nurture theory can also sound judgmental. If Kara is aging rapidly, they suggest, it's all her fault.

Which is right? Nature or nurture? Genes or environment? Actually, both are critical, and it's the interaction between the two that matters most. The real differences between Lisa's and Kara's rates of aging lie in the complex interactions between genes, social relationships and environments, lifestyles, those twists of fate, and especially how one responds to the twists of fate. You're born with a particular set of genes, but the way you live can influence how your genes express themselves. In some cases, lifestyle factors can turn genes on or shut them off. As the obesity researcher George Bray has said, "Genes load the gun, and environment pulls the trigger."[1] His words apply not just to weight gain but to most aspects of health.

We're going to show you a completely different way of thinking about your health. We are going to take your health down to the cellular level, to show you what premature cellular aging looks like and what kind of havoc it wreaks on your body—and we'll also show you not only how to avoid it but also how to reverse it. We'll dive deep into the genetic heart of the cell, into the chromosomes. This is where you'll find **telomeres (tee-lo-meres)**, repeating segments of noncoding DNA that live at the ends of your chromosomes. Telomeres, which shorten with each cell division, help determine how fast your cells age and when they die, depending on how quickly they wear down. The extraordinary discovery from our research labs and other research labs around the world is that the ends of our chromosomes can actually lengthen—and as a result, aging is a dynamic process that can be accelerated or slowed, and in some aspects even reversed. Aging need not be, as thought for so long, a one-way slippery slope toward infirmity and decay. We all will get older, but how we age is very much dependent on our cellular health.

We are a molecular biologist (Liz) and a health psychologist (Elissa). Liz has devoted her entire professional life to investigating telomeres, and her fundamental research has given birth to an entirely new field of scientific understanding. Elissa's lifelong work has been on psychological stress. She has studied its harmful effects on behavior, physiology, and health, and she has also studied how to reverse these effects. We joined forces in research fifteen years ago, and the studies that we performed together have set in motion a whole new way of examining the relationship between the human

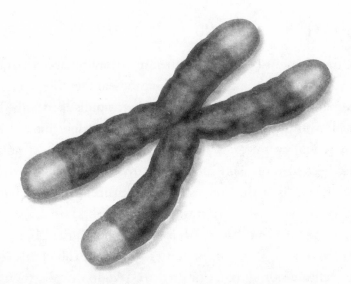

Figure 2: Telomeres at the Tips of Chromosomes. The DNA of every chromosome has end regions consisting of DNA strands coated by a dedicated protective sheath of proteins. These are shown here as the lighter regions at the end of the chromosome—the telomeres. In this picture the telomeres are not drawn to scale, because they make up less than one-ten-thousandth of the total DNA of our cells. They are a small but vitally important part of the chromosome.

mind and body. To an extent that has surprised us and the rest of the scientific community, telomeres do not simply carry out the commands issued by your genetic code. Your telomeres, it turns out, are listening to you. They absorb the instructions you give them. The way you live can, in effect, tell your telomeres to speed up the process of cellular aging. But it can also do the opposite. The foods you eat, your response to emotional challenges, the amount of exercise you get, whether you were exposed to childhood stress, and even the level of trust and safety in your neighborhood—all of these factors and more appear to influence your telomeres and can prevent premature aging at the cellular level. In short, one of the keys to a long healthspan is simply doing your part to foster healthy cell renewal.

HEALTHY CELL RENEWAL AND WHY YOU NEED IT

In 1961 the biologist Leonard Hayflick discovered that normal human cells can divide a finite number of times before they die. Cells reproduce by making copies of themselves (called mitosis), and as the human cells sat in a thin, transparent layer in the flasks that filled Hayflick's lab, they would, at first, copy themselves rapidly. As they multiplied, Hayflick needed more and more flasks to contain the growing cell cultures. The cells in this early stage multiplied so quickly that it was impossible to save all the cultures; otherwise, as Hayflick remembers, he and his assistant would have been "driven out of the laboratory and the research building by culture bottles." Hayflick called this youthful phase of cell division "luxuriant growth." After a while, though, the reproducing cells in Hayflick's lab stopped in their tracks, as if they were getting tired. The longest-lasting cells managed about fifty cell divisions, although most divided far fewer times. Eventually these tired cells reached a stage he called **senescence**: They were still alive but they had all stopped dividing, permanently. This is called the Hayflick limit, the natural limit that human cells have for dividing, and the stop switch happens to be telomeres that have become critically short.

Are all cells subject to this Hayflick limit? No. Throughout our bodies we find cells that renew—including immune cells; bone cells; gut, lung, and liver cells; skin and hair cells; pancreatic cells; and the cells that line our cardiovascular systems. They need to divide over and over and over to keep our bodies healthy. Renewing cells include some types of normal cells that can divide, like immune cells; progenitor cells, which can keep dividing even longer; and those critical cells in our bodies called stem cells, which can divide indefinitely as long as they are healthy. And, unlike those cells in Hayflick's lab dishes, cells don't always have a Hayflick limit, because—as you will read in chapter 1—they have telomerase. Stem cells, if kept healthy, have enough telomerase to enable them to keep dividing throughout our life spans. That cell replenishment, that *luxuriant growth*, is one reason Lisa's skin looks so fresh. It's why her joints move easily. It's one reason she can take in deep lungfuls of the cool air blowing in off the bay. The new cells are constantly renewing essential body tissues and organs. Cell renewal helps keep her feeling young.

From a linguistic perspective, the word *senescent* has a shared history with the word *senile*. In a way, that's what these cells are—they're senile. In one way it is definitely good that cells stop dividing. If they just keep on multiplying, cancer can ensue. But these senile cells are not harmless—they are bewildered and weary. They get their signals confused, and they don't send the right messages to other cells. They can't do their jobs as well as they used to. They sicken. The time of luxuriant growth is over, at least for them. And this has profound health consequences for you. When too many of your cells are senescent, your body's tissues start to age. For example, when you have too many senescent cells in the walls of your blood vessels, your arteries stiffen and you are more likely to have a heart attack. When the infection-fighting immune cells in your bloodstream can't tell when a virus is nearby because they are senescent, you are more susceptible to catching the flu or pneumonia. **Senescent cells can leak proinflammatory substances that make you vulnerable to more pain, more chronic illness. Eventually, many senescent cells will undergo a preprogrammed death.**

The diseasespan begins.

Many healthy human cells can divide repeatedly, so long as their telomeres (and other crucial building blocks of cells like proteins) remain functional. After that, the cells become senescent. Eventually, senescence can even happen to our amazing stem cells. This limit on cells dividing is one reason that there seems to be a natural winding down of the human healthspan as we age into our seventies and eighties, although of course many people live healthy lives much longer. A good healthspan and life span, reaching eighty to one hundred years for some of us and many of our children, is within our reach.[2] There are around three hundred thousand centenarians worldwide, and their numbers are rapidly increasing. Even more so are the numbers of people living into their nineties. Based on trends, it is thought that over one-third of children born in the United Kingdom now will live to one hundred years.[3] How many of those years will be darkened by diseasespan? If we better understand the levers on good cell renewal, we can have joints that move fluidly, lungs that breathe easily, immune cells that fiercely fight infections, a heart that keeps pumping blood through its four chambers, and a brain that is sharp throughout the elderly years.

But sometimes cells don't make it through all their divisions in the way they should. Sometimes they stop dividing earlier, falling into an old, senescent stage before their time. When this happens, you don't get those eight or nine great decades. Instead, you get premature cellular aging. Premature cellular aging is what happens to people like Kara, whose healthspan graph turns dark at an early age.

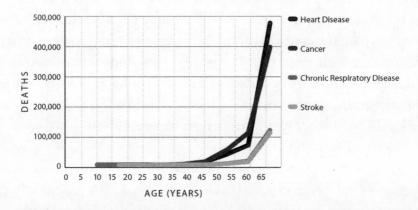

Figure 3: Aging and Disease. Age is by far the largest determinant of chronic diseases. This graph shows the frequency of death by age, up to age sixty-five and older, for the top four causes of death by disease (heart disease, cancer, respiratory disease, and stroke and other cerebrovascular diseases). The death rate due to chronic diseases starts to increase after age forty and goes up dramatically after age sixty. Adapted from U.S. Department of Health and Human Services, Centers for Disease Control and Prevention, "Ten Leading Causes of Death and Injury," http://www.cdc.gov/injury/wisqars/leadingCauses.html.

Chronological age is the major determinant of when we get diseases, and this reflects our biological aging inside.

At the beginning of the chapter, we asked, *Why do people age differently?* One reason is cellular aging. Now the question becomes, *What causes cells to get old before their time?*

For an answer to this question, think of shoelaces.

HOW TELOMERES CAN MAKE YOU FEEL OLD OR HELP YOU STAY YOUNG AND HEALTHY

Do you know the protective plastic tips at the ends of shoelaces? These are called aglets. The aglets are there to keep shoelaces from fraying. Now imagine that your shoelaces are your chromosomes, the structures inside your cells that carry your genetic information. Telomeres, which can be measured in units of DNA known as base pairs, are like the aglets; they form little caps at the ends of the chromosomes and keep the genetic material from unraveling. They are the aglets of aging. But telomeres tend to shorten over time.

Here's a typical trajectory for the life of a human's telomere:

Age	Telomere Length (in base pairs)
Newborn baby	10,000 base pairs
35 years old	7,500 base pairs
65 years old	4,800 base pairs

When your shoelace tips wear down too far, the shoelaces become unusable. You may as well throw them away. Something similar happens to cells. When telomeres become too short, the cell stops dividing altogether. Telomeres aren't the only reason a cell can become senescent. There are other stresses on normal cells that we don't yet understand very well. But short telomeres are one of the primary reasons human cells grow old, and they are one mechanism that controls the Hayflick limit.

Your genes affect your telomeres, both their length when you're born and how quickly they dwindle down. But the wonderful news is that our research, along with research from around the globe, has shown you can step in and take some control of how short or long—how *robust*—they are.

For instance:

- Some of us respond to difficult situations by feeling highly threatened—and this response is linked to shorter telomeres. We can reframe our view of situations in a more positive way.
- Several mind-body techniques, including meditation and Qigong, have been shown to reduce stress *and* to increase telomerase, the enzyme that replenishes telomeres.
- Exercise that promotes cardiovascular fitness is great for telomeres. We describe two simple workout programs that have been shown to improve telomere maintenance, and these programs can accommodate all fitness levels.
- Telomeres hate processed meats like hot dogs, but fresh, whole foods are good for them.
- Neighborhoods that are low in social cohesion—meaning that people don't know and trust one another—are bad for telomeres. This is true no matter what the income level.
- Children who are exposed to several adverse life events have shorter telomeres. Moving children away from neglectful circumstances (such as the notorious Romanian orphanages) can reverse some of the damage.
- Telomeres on the parents' chromosomes in the egg and sperm are directly transmitted to the developing baby. Remarkably, this means that if your parents had hard lives that shortened their telomeres, they could have passed those shortened telomeres on to you! If you think that might be the case, don't panic. Telomeres can build up as well as shorten. **You can still take action to keep your telomeres stable. And this**

news also means that our own life choices can result in a positive cellular legacy for the next generation.

MAKE THE TELOMERE CONNECTION

When you think about living in a healthier way, you may think, with a groan, about a long list of things you ought to be doing. For some people, though, when they have seen and understood the connection between their actions and their telomeres, they are able to make changes that last. When I (Liz) walk to the office, people sometimes stop me to say, "Look, I'm biking to work now—I'm keeping my telomeres long!" Or "I stopped drinking sugary soda. I hated to think of what it was doing to my telomeres."

WHAT'S AHEAD

Does our research show that by maintaining your telomeres you will live into your hundreds, or run marathons when you're ninety-four, or stay wrinkle free? No. Everyone's cells become old and eventually we die. But imagine that you're driving on a highway. There are fast lanes, there are slow lanes, and there are lanes in between. You can drive in the fast lane, barreling toward the diseasespan at an accelerated pace. Or you can drive in a slower lane, taking more time to enjoy the weather, the music, and the company in the passenger seat. And, of course, you'll enjoy your good health.

Even if you are currently on a fast track to premature cellular aging, you can switch lanes. In the pages ahead, you'll see how to make this happen. In the first part of the book, we'll explain more about the dangers of premature cellular aging—and how healthy telomeres are a secret weapon against this enemy. We'll also tell you about the discovery of telomerase, an enzyme in our cells that helps keep the protective sheaths around our chromosome ends in good shape.

The rest of the book shows you how to use telomere science to support your cells. Begin with changes that you can make to your mental habits and then to your body—to the kinds of exercise, food, and sleep routines that are best for telomeres. Then expand outward to determine whether your social and physical environments support your telomere health. Throughout the book, sections called "Renewal Labs" offer suggestions that can help you prevent premature cellular aging, along with an explanation of the science behind those suggestions.

By cultivating your telomeres, you can optimize your chances of living a life that is not just longer but better. That is, in fact, why we've written this book. In the course of our work on telomeres we've seen too many Karas—too many men and women whose telomeres are wearing down too fast, who enter the diseasespan when they should still feel vibrant. There is abundant high-quality research, published in prestigious scientific journals and backed by the best labs and universities, that can guide you toward avoiding this fate. We could wait for those studies to trickle down through the media and make their way into magazines and onto health websites, but that process can take many years and is piecemeal, and, sadly, information often gets distorted along the way. We want to share what we know now—and

THE HOLY GRAIL?

Telomeres are an integrative index of many lifetime influences, both the good, restorative ones like good fitness and sleep, and also malign ones like toxic stress or poor nutrition or adversities. Birds, fish, and mice also show the stress-telomere relationship. Thus it's been suggested that telomere length may be the "Holy Grail for cumulative welfare,"[4] to be used as a summative measure of the animals' lifetime experiences. In humans, as in animals, while there will be no one biological indicator of cumulative lifetime experience, telomeres are among one of the most helpful indicators that we know of right now.

we don't want more people or their families to suffer the consequences of unnecessary premature cellular aging.

When we lose people to poor health, we lose a precious resource. Poor health often saps your mental and physical ability to live as you wish. When people in their thirties, forties, fifties, sixties, and beyond are healthier, they will enjoy themselves more and will share their gifts. They can more easily use their time in meaningful ways—to nurture and educate the next generation, to support other people, solve social problems, develop as artists, make scientific or technological discoveries, travel and share their experiences, grow businesses, or serve as wise leaders. As you read this book, you are going to learn a lot more about how to keep your cells healthy. We hope you're going to enjoy hearing how easy it is to extend your healthspan. And we hope you're going to enjoy asking yourself the question: *How am I going to use all those wonderful years of good health?* Follow a bit of the advice in this book, and chances are that you'll have plenty of time, energy, and vitality to come up with an answer.

RENEWAL BEGINS RIGHT NOW

You can start to renew your telomeres, and your cells, right now. One study has found that people who tend to focus their minds more on what they are currently doing have longer telomeres than people whose minds tend to wander more.[5] Other studies find that taking a class that offers training in mindfulness or meditation is linked to improved telomere maintenance.[6]

Mental focus is a skill that you can cultivate. All it takes is practice. You'll see a shoelace icon, pictured here, throughout the book. Whenever you see it—or whenever you see your own shoes with or without laces—you might use it as a cue to pause and ask yourself what you're thinking. Where are your thoughts right now? If you're worrying or rehashing old problems, gently remind yourself

15

to focus on whatever it is you're doing. And if you are not "doing" anything at all, then you can enjoy focusing on "being."

Simply focus on your breath, bringing all of your awareness to this simple action of breathing in and out. It is restorative to focus your mind inside (noticing sensations, your rhythmic breathing), or outside (noticing the sights and sounds around you). This ability to focus on your breath, or your present experience, turns out to be very good for the cells of your body.

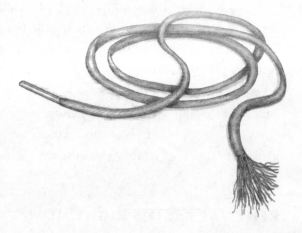

Figure 4: Think of Your Shoelaces. Shoelace tips are a metaphor for telomeres. The longer the protective aglets at the ends of the laces, the less likely the shoelace will fray. In terms of chromosomes, the longer the telomeres, the less likely there will be any alarms going off in cells or fusions of chromosomes. Fusions trigger chromosome instability and DNA breakage, which are catastrophic events for the cell.

Throughout the book, you will see a shoelace icon with long aglets. You can use that as an opportunity to refocus your mind on the present, take a deep breath, and think of your telomeres being restored with the vitality of your breath.

PART I

TELOMERES:
A PATHWAY TO LIVING
YOUNGER

CHAPTER ONE

How Prematurely Aging Cells Make You Look, Feel, and Act Old

Ask yourself these questions:

1. How old do I look?

- ▪ I look younger than my age.
- ▪ I look about my age.
- ▪ I look older than my age.

2. How would I rate my physical health?

- ▪ I'm in better health than most people my age.
- ▪ I'm about as healthy as most people my age.
- ▪ I'm less healthy than most people my age.

3. How old do I feel?

- ▪ I feel younger than my age.
- ▪ I feel about my age.
- ▪ I feel older than my age.

These three questions are simple, but your answers can reveal important trends in your health and aging. People who look older than their age can in fact be experiencing the early hair graying or skin damage associated with shorter telomeres. Poor physical health can happen for lots of reasons, but an early entry into the diseasespan is often a sign that your cells are aging. And studies show that people who feel older than their biological age also tend to be sicker earlier than people who feel younger.

When people say that they fear getting older, what they usually mean is that they fear a long, drawn-out diseasespan. They fear trouble getting up the stairs, trouble recovering from open-heart surgery, trouble wheeling around a tank of oxygen; they fear loss of bone, curved backs, and the dreaded loss of memory and of mind. And they fear a consequence of all these: loss of opportunities for healthy social connections and the need to replace those with dependency on others. But really, aging doesn't have to be so traumatic.

If your answers to our three questions suggest that you look and feel older than your age, perhaps it's because your telomeres are wearing down faster than they should. Those short telomeres could be sending your cells a signal that it's time to fast-forward the aging process. It's an alarming scenario, but take heart. There's plenty you can do to fight premature aging where it counts the most: at the cellular level.

However, you can't successfully fight your enemy until you really understand it.

In this section of the book, we'll give you the knowledge you need before you begin the battle. This first chapter scouts out what happens during premature cellular aging. You'll take a close-up look at aging cells and see why they are so damaging to your body and brain. You'll also discover why many of the most frightening and debilitating diseases are linked to short telomeres and thus cell aging. Then, in chapters 2 and 3, you'll see how telomeres and the fascinating enzyme telomerase (pronounced *tell-OMM-er-ase*) can either trigger an early diseasespan or work to keep your cells healthy.

HOW ARE PREMATURELY AGING CELLS DIFFERENT FROM HEALTHY CELLS?

Think of the human body as a barrel full of apples. A healthy human cell is like one of these fresh, shiny apples. But what if there is a rotten apple in that barrel? Not only can't you eat it, but worse, it will start to make the other apples around it rotten, too. That rotten apple is like an aged, senescent cell in your body.

Before we explain why, we want to return to the fact that your body is full of cells that need to constantly renew themselves to stay healthy. These renewing cells, which are called proliferative cells, live in places like your:

- immune system
- gut
- bones
- lungs
- liver
- skin
- hair follicles
- pancreas
- cardiovascular system lining
- heart's smooth muscle cells
- brain, in parts including the hippocampus (a learning and memory center of the brain)

For these crucial body tissues to stay healthy, their cells need to keep renewing. Your body has finely calibrated systems for assessing when a cell needs to be renewed; even though a body tissue can look the same for years, it is constantly being replaced by new cells in just the right numbers and at the correct rate. But remember that some cells have a limit to how many times they can divide. When cells can no longer renew themselves, the body tissues they supply will start to age and function poorly.

Cells in our tissues originate from stem cells, which have the amazing ability to become many different types of specialized cells. They live in stem cell niches, which are a kind of VIP lounge where stem cells are protected and lie dormant until they are needed. The niches are usually in or near the tissues that the stem cells will replace. Stem cells for skin live under the hair follicles, some stem cells for the heart live in the right ventricular wall, and muscle stem cells live deep in the fiber of the muscle. If all is well, the stem cells remain in their niche. But when there is a need to replenish tissues, the stem cell appears on deck. It divides and produces proliferative cells—sometimes called progenitor cells—and some of their progeny cells transform into whatever specialized cell is needed. If you get sick and need more immune cells (white blood cells), freshly divided stem cells for blood that were hiding out in the bone marrow will enter the bloodstream. Your gut lining is constantly being worn down by normal digestive processes, and your skin is being sloughed off, and stem cells keep these body tissues replenished. If you go jogging and tear your calf muscle, some of your muscle stem cells will divide, each stem cell creating two new cells. One of those cells replaces the original stem cell and remains comfortably in its niche; the other can become a muscle cell and help replenish the damaged tissue. Having a good supply of stem cells that are able to renew themselves is key to staying healthy and to recovering from sickness and injury.

But when a cell's telomeres become too short, they send out signals that put the cell's cycle of division and replication under arrest. An arrested cell stops in its tracks. The cell can no longer renew itself. It becomes old; it becomes senescent. If it is a stem cell, it goes into permanent retirement and will no longer leave its cozy niche when it is needed. Other cells that become senescent just sit around, unable to do the things they're supposed to do. Their internal powerhouses, the mitochondria, don't work properly, causing a kind of energy crisis.

An old cell's DNA can't communicate well with the other parts of the cell, and the cell can't keep house well. The old cell gets crowded inside, with—among other things—clumps of malfunctioning proteins and brown globs of "junk" known as lipofuscin, which can cause macular degeneration in the eyes and some neurological diseases. Worse still—and why they are like rotten apples in a barrel—senescent cells send out false alarms in the form of pro-inflammatory substances, reaching other parts of the body as well.

The same basic process of aging happens across the different types of cells in our bodies, whether they are liver cells, skin cells, hair follicle cells, or the cells that line our blood vessels. But there are some twists on the process that depend on the cell's type and its location in the body. Senescent cells in the bone marrow prevent blood and immune stem cells from dividing the way they're supposed to, or warp them into making blood cells in unbalanced amounts. Senescent cells in the pancreas may not correctly "hear" signals that regulate their production of insulin. Senescent cells in the brain may secrete substances that cause neurons to die. While the underlying process of aging is similar in most of the cells that have been studied, a cell's way of expressing that aging process can create different kinds of injury to the body.

Aging can be defined as the cell's "progressive functional impairment and reduced capacity to respond appropriately to environmental stimuli and injuries." Aged cells can no longer respond to stresses normally, whether the stress is physical or psychological.[1] This process is a continuum that often silently and slowly segues into the diseases of aging—diseases that can be traced, in part, to shorter telomeres and aging cells. To understand aging and telomeres a little better, let's go back to the three questions that we asked you at the beginning of this chapter:

How old do you look?

How would you rate your physical health?

How old do you feel?

OUT WITH THE OLD, IN WITH THE NEW: REMOVING SENESCENT CELLS IN MICE REVERSES PREMATURE AGING

One laboratory study followed mice that had been genetically altered so that many of their cells were senescent much earlier than usual. The mice began to age prematurely—they lost fat deposits, which made them look wrinkled; their muscles withered; their hearts weakened; and they developed cataracts. Some died early of heart failure. Then, in an experimental genetic trick that is not possible to replicate in humans, the researchers removed the mice's senescent cells. Taking out the senescent cells reversed many of the symptoms of premature aging. It cleaned up their cataracts and restored their wasted muscles, maintained their fat deposits (which reduced their wrinkles), and promoted a longer healthspan.[2] **Senescent cells control the aging process!**

PREMATURELY AGING CELLS: HOW OLD DO YOU LOOK?

Age spots and blotches. Gray hair. The shrunken or stooped posture that comes with bone loss. These changes happen to all of us, but if you've been to a high school reunion recently, you've seen proof that they don't happen at the same time or in the same way.

Walk through the doors of your tenth high school reunion, when everyone is still in their twenties, and you'll spot classmates who sport expensive clothes—and classmates whose party outfits look a bit threadbare. Some classmates are parading their career successes, start-up companies, or productivity in offspring, and others are gulping down Scotch while commiserating about their latest heartbreaks. It may not seem fair. But in terms of the physical signs of aging, it's a level field. Almost everyone in the room—no matter whether they're rich, poor, successful, struggling, happy, or sad—will *look* like they're in their twenties. Their hair is healthy, their skin is clear, and a few of your classmates are an inch or two taller than when your class graduated ten years ago. They are in the radiant peak of young adulthood.

But show up for a reunion five or ten years later, and a different scene emerges. You'll notice that a few of your old classmates are starting to look like *old* classmates. They're a little gray around the ears or showing more forehead. Their skin looks speckled and cloudier; the crow's-feet around their eyes may be etched deeply. They may have protruding bellies and may even look a bit hunched over. These people are experiencing a rapid onset of outward physical aging.

Yet other classmates are graced with a slower aging trajectory. Over the years, as your twentieth, thirtieth, fortieth, fiftieth, and sixtieth reunions come and go, it's evident that the hair, faces, and bodies of these lucky classmates are changing—but these changes happen slowly and gradually, with elegance. Telomeres, as you'll see, play at least a small role in how quickly you develop an aged appearance, and whether you become one of those people who "age well."

Skin Aging

The skin's outer layer, the epidermis, is made up of proliferating cells that are constantly replenishing themselves. Some of these skin cells (the keratinocytes) make telomerase, so they don't wear out and become senescent cells, but most do slow down in their ability to replenish.[3] Underneath this visible layer of skin is the dermis, a layer of skin cells (skin fibroblasts) that creates the foundation for a healthy, plump epidermis—by producing collagen, elastin, and growth-promoting factors, for example.

With age, these fibroblasts secrete less collagen and elastin, which makes the outer layer of visible skin look old and loose. This effect translates upward through the skin layers to create a more aged outer appearance. Aged skin becomes thinner, as it loses fat pads and hyaluronic acid (which acts as a natural moisturizer for skin and joints). It becomes more permeable to the elements.[4] The aged melanocytes lead to age spots but also paleness. In short, the aging skin gets the all-too-familiar spotty, pale, saggy, and wrinkly look, mainly due to the aging fibroblasts that no longer support the outer cells.

In older people, skin cells often lose their ability to divide. Some older people do have skin cells that can still keep dividing. When researchers peer into their cells, they see the cells are better at fending off oxidative stress and have longer telomeres.[5] Although short telomeres don't necessarily cause aging skin, they play a role, especially when it comes to aging from the sun (also called photoaging). The UV rays from sun exposure can damage telomeres.[6] Petra Boukamp, a telomere skin researcher from the German Cancer Research Center in Heidelberg, and her colleagues have compared skin from a sun-exposed site—the neck—to a sun-protected site—the buttocks. The outer cells on the neck showed some telomere attrition from the sun, whereas the protected buttock cells showed almost no telomere attrition with aging! Skin cells, when protected from the sun, can withstand aging for a long time.

Bone Loss

Your bone tissue is remodeled throughout your life, and a healthy level of bone density results when you have a balance between the bone-building cells (osteoblasts) and the bone-busting cells (osteoclasts). Osteoblast cells need healthy telomeres in order to keep dividing and replenishing themselves—and when your telomeres are short, the osteoblasts get old and can't keep up with the osteoclasts. The balance tips, and the osteoclasts nibble away at your bones.[7] It doesn't help that after a person's telomeres wear down, the old bone cells become inflammatory. Lab mice bred to have extra-short telomeres suffer from early bone loss and osteoporosis;[8] so do people who have been born with a genetic disorder that causes their telomeres to be extraordinarily short.

Graying Hair

In a sense, we're all born with hair that's been colored. Each strand of hair begins inside its own follicle and is made from keratin, which produces a white hair. But there are special cells inside the follicle— melanocytes, the same kinds of cells responsible for skin color—that inject the hair with pigment. Without these natural hair-dye cells, hair color is lost. Stem cells in the follicle produce the melanocytes. When these stem cells' telomeres wear down, the cells can't replenish themselves fast enough to keep up with hair growth, and gray hair is a result. Eventually, when all the melanocytes have died, hair becomes pure white. Melanocytes are also sensitive to chemical stressors and to ultraviolet radiation; and in a study published in the journal *Cell*, mice who underwent X-rays developed damaged melanocytes and gray fur.[9] Mice with a genetic mutation that causes extremely short telomeres also develop early graying of their fur, and restoring telomerase turns their gray fur dark again.[10]

What's normal graying? Graying happens least in African Americans and Asians, and most in blonds.[11] Graying starts to happen in at least half of people by their late forties, and around 90 percent of people in their early sixties. The vast majority of cases of early graying are quite normal; only a very few people who find themselves with gray or white hair at an early age, in their thirties, may carry a genetic mutation that causes short telomeres.

What Does Your Appearance Say About Your Health?

Maybe you're thinking, "Well, I don't really mind having a few early gray hairs. And are a couple of age spots around my eyes really such a big deal? Aren't you asking me to focus on the wrong things—to value a youthful appearance instead of my health?" These are great questions. There is no contest here: health is what matters. But how much does aged appearance reflect inner health? Researchers have asked specially trained "raters" to estimate the age of a person just by looking at a photo.[12] As it turns out, the people who look older on average have shorter telomeres. This is not surprising, given the role that telomeres appear to play in skin aging and hair graying. Looking older is associated in small but worrisome ways with signs of poor physical health. People who look old tend to be weaker, to perform worse on a mental exam that tests memory, to have higher fasting glucose and cortisol levels, and to show early signs of cardiovascular disease.[13] The good news is that these are *very small effects*. What's on the inside of your body is what matters most, but looking older than your age—looking haggard—is a sign worth paying attention to. It may be an indicator that your telomeres need more protection.

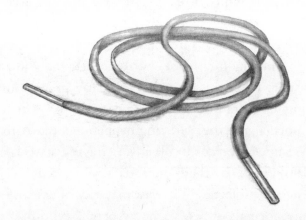

Remember what to do when you see this image? See page 16.

PREMATURE CELLULAR AGING: HOW IS YOUR PHYSICAL HEALTH?

You can see the real power of short telomeres to damage your cells and your health when you consider the next question: *How would you rate your physical health?*

Think again about your high school reunions. As you reach your twentieth or thirtieth reunion, you'll notice that many of your classmates are starting to suffer from the common diseases of aging. Yet they're only about forty or fifty. They're not chronologically old yet. So why are their bodies *acting* as if they are? Why are they entering the diseasespan at young ages?

Inflamm-Aging

Wouldn't it be interesting if you could peer into the cells deep inside each person at your reunion and measure the lengths of each person's telomeres? If you could, you would see that the people with the shortest telomeres are on average the ones who are sicker, weaker, or whose faces show the strain of coping with health problems like diabetes, cardiovascular disease, a weakened immune system, and lung diseases. You would probably also find that the ones with the shortest telomeres suffer from chronic inflammation. The observation that inflammation increases with age and is a cause of the diseases of aging is so important that scientists have a name for it: *inflamm-aging*. This is a persistent, low-grade inflammation that can accumulate with age. There are many reasons why this occurs, such as proteins becoming damaged. One other common cause of inflamm-aging involves telomere damage.

When a cell's genes are damaged or its telomeres are too short, that cell knows its precious DNA is in danger. The cell reprograms itself so that it emits molecules that can travel to other cells and call for help. These molecules, together called senescence-associated secretory phenotype, or SASP, can be useful. If a cell has become senescent because it's been wounded, it can send signals to neighboring immune cells and other cells with repair functions, to call in the squads that can get the healing process going.

And here's where things go terribly wrong. Telomeres have an abnormal response to DNA damage. The telomere is so preoccupied with protecting itself that even though the cell has called out for help, the telomere won't let the help in. It's like people who doggedly refuse assistance in the face of adversity because they're afraid to let their guard down. A shortened telomere can sit inside an aging cell for months, signaling and signaling for help but not allowing the cell to take action to resolve the damage. This unremitting but futile signaling can have devastating consequences. Because now that cell becomes like the rotten apple in the barrel. It starts affecting all the tissues around it. The SASP process involves chemicals like proinflammatory cytokines that, over time, travel through the body, leading to system-wide chronic inflammation. Judith Campisi of the Buck Institute of Aging discovered SASP, and she has shown that these cells create a friendly territory for cancer growth.

In the past decade or so, scientists have come to recognize that chronic inflammation (from SASP or other sources) is a crucial player in causing many diseases. Short-term, acute inflammation brings healing to injured cells, but long-term inflammation interferes with the normal functioning of body tissues. For example, chronic inflammation can cause cells of the pancreas to malfunction and not regulate insulin production properly, setting the stage for diabetes. It

Figure 5: A Rotten Apple in an Apple Barrel. Think of a barrel of apples. An apple barrel's health depends on each apple. One rotten apple sends out gases that rot the other apples. One senescent cell sends signals to surrounding cells, promoting inflammation and factors that promote what we might call "cell rot."

can cause plaque on an artery wall to burst. It can cause the body's immune response to turn on itself, so that it attacks its own tissues.

These are just a few of the most harrowing examples of inflammation's destructive power, but the list marches on to a deadly drumbeat. Chronic inflammation is also a factor in heart disease, brain diseases, gum disease, Crohn's disease, celiac disease, rheumatoid arthritis, asthma, hepatitis, cancers, and more. That's why scientists talk about inflamm-aging. It's real.

If you want to slow inflamm-aging, if you want to stay in the healthspan for as long as possible, you've got to prevent chronic inflammation. And a big part of controlling inflammation means protecting your telomeres. Since cells with very short telomeres

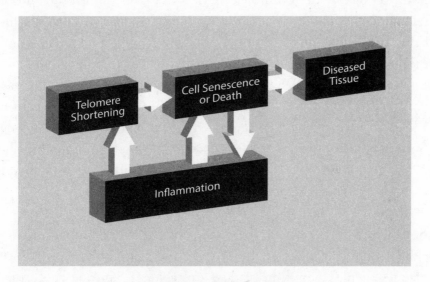

Figure 6: A Path from Short Telomeres to Disease. One early pathway to disease is telomere shortening. Shortened telomeres lead to senescent cells that either stick around or—if we are lucky—are removed from the scene early. While there are many factors that can cause senescence, telomere damage is a common one in humans. When the old senescent cells build up over decades to a critical mass, they become the foundation for diseased tissue. Inflammation is a cause of both telomere shortening and senescent cells, and senescent cells in turn create more inflammation.

send out constant inflammatory signals, you need to keep those telomeres a healthy length.

Heart Disease and Short Telomeres

Each of our arteries, from the large to the small, is lined with layers of cells called the endothelium. If you want your cardiovascular system to stay healthy, the cells of the endothelium need to replenish themselves, protect the lining, and keep immune cells from entering the arterial wall.

But in people with short telomeres in their white blood cells, the risk of cardiovascular disease goes up. (Usually, short telomeres in the blood reveal short telomeres in other tissues, like the endothelium.) People with common genetic variations that lead to shorter telomeres are also more prone to cardiovascular problems.[14] Just being in the bottom third of the population's blood telomere length means you are 40 percent more likely to develop cardiovascular disease in the future.[15] Why? We don't know all the pathways, but vascular senescence is one of them: When short telomeres tell cells to age prematurely, the endothelium can't renew itself to make strong, smooth blood vessel linings. It becomes weaker and more vulnerable to disease. When the actual vascular tissue with plaques is examined, short telomeres are indeed found.

In addition, short blood cell telomeres can also trigger inflammation, which sets the stage for cardiovascular disease. Inflammatory cells stick to the sides of the arteries and trap cholesterol to form plaques or make existing plaques unstable. If a plaque ruptures, a blood clot can form over the plaque, blocking the artery. And if this artery is a coronary artery, it chokes off the heart's blood supply and causes a heart attack.

Lung Diseases and Short Telomeres

People with asthma, chronic obstructive pulmonary disease (COPD), and pulmonary fibrosis (a very serious, irreversible disease in which scarred lung tissue leads to difficult breathing) have shorter telomeres in their immune cells and lung cells than people who are healthy. Pulmonary fibrosis in particular clearly results from not having good telomere maintenance; the proof for this comes from finding pulmonary fibrosis in those unlucky people with rare inherited telomere maintenance gene mutations. Along with this revealing fact, several other lines of smoking gun evidence exist. Together they point a strongly incriminating finger to inadequate telomere maintenance as a shared underlying problem that contributes to COPD, asthma, lung infections, and poor lung functioning—and this is true in everyone, not just people with a rare mutation in a telomere maintenance gene. Lacking robust telomere maintenance, lung stem cells and blood vessels of the lungs become senescent. They cannot keep the lung tissues replenished and supplied with their needs. Immune-cell senescence creates a proinflammatory environment that further taxes the lungs, so they function more and more poorly.

PREMATURE CELLULAR AGING: HOW OLD DO YOU FEEL?

Let's return to your high school reunion—this time, let's go to your fortieth reunion, when your classmates are closing in on age sixty. This is when the first people in your class start to show signs of cognitive slowing. It may be hard to put your finger on what exactly is different about these folks, but you'll notice that they seem a little fuzzier, a little out of it, a little less focused, and less tuned in to normal social cues. They may take a few extra seconds to remember your name. This mental loss, more than anything else, is what makes us feel really, truly old.

Cognitive Slowing and Alzheimer's Disease

You won't be surprised to hear that the people who have early cognitive problems also tend to have shorter telomeres. This effect may persist as people get older. In one study of otherwise healthy seventy-year-olds, shorter telomeres predicted general cognitive decline years later.[16] In young adults, there was no relationship between telomeres and cognitive function, but greater telomere shortening over approximately ten years predicted poorer cognitive function.[17] Researchers are fascinated by a possible relationship between the lengths of our telomeres and the sharpness of our thinking. Can short telomeres predict dementia or Alzheimer's disease?

A large, impressive Texas study set out to help answer this question.[18] Researchers imaged the brains of almost two thousand adults from Dallas County. The study controlled for age as well as other factors that affect the brain, such as smoking, gender, and the status of a gene, APOE-epsilon 4 (commonly called just APOE). A normal variant of APOE increases a person's risk of Alzheimer's. As expected, nearly everyone's brain showed some signs of shrinking with age. But then the researchers examined the parts of the brain that are specifically involved in emotions and memory. The hippocampus, for example, is a part of the brain that helps form, organize, and store memories; it also helps link those memories to your emotions and your senses. The hippocampus is a reason that the scent of a new box of erasers drops you back into the first day of grade school; it's a reason you can remember grade school at all. Amazingly, the Texas researchers found that when people had short telomeres in their white blood cells (which serve as a window into telomere length throughout the body), their hippocampuses were smaller than those of people with longer telomeres. The hippocampus is made up of cells that need to regenerate—and if you want to have good memory function, it's essential for your body to be able to replenish the hippocampus's cells.

It isn't just the hippocampus that is smaller in people with short telomeres. So are other areas of the brain's limbic system, including the amygdala and the temporal and parietal lobes. These areas, along with the hippocampus, help regulate memory, emotion, and stress—*and* they are the very same areas that atrophy in Alzheimer's disease. The Dallas study suggests that *short telomeres in the blood crudely indicate an aging brain.* It is possible that cellular aging, perhaps just in the hippocampus or perhaps throughout the body, may underlie an important pathway of dementia. Keeping telomeres healthy may be especially crucial for people who carry the variant of APOE gene that puts them at a higher risk for early Alzheimer's. One study found if you have this gene variant and also have short telomeres, your risk of dying earlier is nine times greater than if you have the same gene variant but your telomeres are long.[19]

Short telomeres may help cause Alzheimer's directly. There are common genetic variations (in genes called TERT and OBFC1) that can lead to short telomeres. Remarkably, people who have even just one gene with these common variations are statistically more likely to develop Alzheimer's.[20] This is not a big effect, but it demonstrates a causal relationship: telomeres are not just a marker for something else, or an epiphenomenon, but rather they are causing part of brain aging—putting us at greater risk for neurodegenerative disease processes. TERT and OBFC1 directly function to maintain telomeres in well-understood ways. The evidence keeps growing. If you want to keep your brain sharp, think about your telomeres. See notes at the back of this book for a research opportunity on brain aging.[21]

A Healthy "Felt Age"

If you went to your fortieth reunion, climbed up onto the stage, and asked this group of sixty-year-olds to raise their hands if they *feel* like sixty-year-olds, you'd get an interesting result. A majority of people—75 percent—would say that they feel younger than their age. Even as the years go by, and even as the date of birth on our driver's license tells us that we are getting older, many of us still *feel* young.[22] This response to aging is highly adaptive. Having a younger "felt age" is associated with more life satisfaction, personal growth, and social connections with others.[23]

Feeling younger is different from wishing to *be* younger. People who long to be a chronologically younger age (say, a man in his fifties who wishes he could be thirty again) tend to be unhappier and more dissatisfied with their lives. Wishing and longing for youth is really the opposite of our major developmental task as we age, which is to accept ourselves as we are, even while working toward maintaining our mental and physical fitness.

FOR A HEALTHIER OLD AGE, CHANGE HOW YOU THINK ABOUT IT

Be careful of how you think about old people. People who internalize and accept negative age stereotypes may *become* age stereotypes—they may develop more health challenges. This phenomenon, called stereotype embodiment, was identified by Becca Levy, a social psychologist at Yale University. Even when their current health status is taken into account, people who have negative beliefs about aging act differently from people who have a sunnier view of aging.[24] They believe they have less control over whether they develop disease, and they don't work as hard at health behaviors, such as taking prescribed medications. They're more than twice as likely to die of a heart attack, and as the decades go by they experience a steeper decline in memory. When they are injured or sick, they recover more slowly.[25] In another study, elderly people who were simply reminded of age stereotypes performed so poorly on a test that they scored as low as if they had dementia.[26]

If you have a negative view of aging, you can make a conscious effort to counter it. Here's a list of stereotypes we've adapted from Levy's Image of Aging Scale.[27] You might visualize yourself thriving in old age, embodying some of those positive traits. When you catch yourself thinking of old age negatively, remind yourself of the positive side of aging:

What's Your Image of Aging?	
grumpy	optimistic
dependent	capable
slow	full of vitality
frail	self-reliant
lonely	strong will to live
confused	wise
nostalgic	emotionally complex
distrustful	close relationships
bitter	loving

What is the profile of our emotional life as we age? Despite the image of older people as cranky or resentful of the young, Laura Carstensen, a researcher of aging at Stanford University, shows that our daily emotional experience is actually enhanced with age. Typically, older people experience more positive emotions than negative ones in daily life. The experience isn't purely "happy." Rather, our emotions grow richer and more complex over time. We experience more co-occurrence of positive and negative emotions, such as those poignant occasions when you get a tear in the eye at the same time you feel joy, or feeling pride at the same time you feel anger[28]—a capacity we call "emotional complexity." These mixed emotional states help us avoid the dramatic ups and downs that younger people have, and they also help us exercise more control over what we feel. Mixed emotions are easier to manage than purely positive or purely negative emotions. Thus, emotionally speaking, life just feels better. Better control over emotions and enhanced complexity means more enriched daily experiences.

People with more emotional complexity also have a longer healthspan.[29]

Gerontology researchers also know that we maintain interest in intimacy and sex as we age. Our social circles get smaller, but this is largely by choice. Over time, we shape our social circles to include the most meaningful relationships, and we weed away those more troublesome relationships. This leads to days with more positive feelings and less stress. We prioritize things better and focus our time on things that matter most to us. Perhaps that's one way to describe the wisdom of age.

Your efforts to imagine a better, healthier, more vibrant old age will pay off. Levy reminded older people of the benefits of aging—like enlightenment and accomplishment—and then gave them stressful tasks to perform. She found that they responded to the stress with less reactivity (lower heart rate and blood pressure) than a control group.[30] As the saying goes, "Age is an issue of mind over matter. If you don't mind, it doesn't matter."

TWO PATHS

Pause for a minute. Imagine what your future might look like if your telomeres shorten too quickly and your cells begin to age prematurely. This thought exercise is intended to make premature cellular aging vivid and real for you. Think about the kind of aging you *don't* want to experience in your forties, fifties, sixties, and seventies. Do you dread scenarios like these?

- "I've lost my sharpness. When I talk, my younger colleague's eyes glaze over because I'm rambling and unfocused."
- "I'm always in bed with a respiratory infection; I seem to catch every illness."

- "It is hard to breathe."
- "My legs feel numb."
- "My feet aren't steady underneath me. I'm scared of falling."
- "I'm too tired to do anything but sit on the couch watching TV all day."
- "I overhear my children saying, 'Whose turn is it to take care of Mom?'"
- "I can't travel the way I'd planned to, because I want to stay close to my doctors."

These statements reveal aspects of life with an early diseasespan—the kind of life you want to avoid. You may have parents or grandparents who believed the old myth that everyone gets a few good decades and then it's time to get sick or give up. We all know people who turned sixty or seventy and quietly declared that their lives were finished. These are the folks who pull on their sweatpants, sit back in the recliner, and watch television until disease takes over.

Now envision a different future, one with long, healthy telomeres and with cells that renew. What do these decades of good health look like? Do you have a role model you can picture?

Aging is often portrayed in such negative ways that most of us try not to even think about it. If you had parents or grandparents who got sick early, or who simply gave up once they hit a certain age, it may be hard for you to imagine that it's possible to be old, healthy, and energetically engaged with life. But if you can form a clear, positive picture of how you'd like to age, you suddenly have a goal to work toward while aging—and a compelling reason to keep your telomeres and cells healthy. If you think of aging in a positive way, odds are that you'll live seven and a half years longer than someone who doesn't, at least according to one study![31]

One of our favorite examples of a person who is constantly renewed in spirit is my (Liz's) friend Marie-Jeanne, a delightful molecular biologist who lives in Paris. Marie-Jeanne is about eighty years old; she has white hair and wrinkles, and her back is slightly stooped, but her face is lively and intelligent. Marie-Jeanne and I met up for the afternoon recently. We had lunch. We visited the Petit Palais art museum, walking upstairs and downstairs and seeking out most of the exhibits. We explored the Latin Quarter on foot and visited bookshops. Six hours later, Marie-Jeanne was looking fresh with no sign of slowing down. I was ready to drop from exhaustion. I proposed heading back ("so Marie-Jeanne could rest"). As Marie-Jeanne suggested yet another venue to visit, I, ashamed to admit how desperate I was to put up my aching feet, conjured up a previous engagement so my tired legs could get home to collapse.

Marie-Jeanne checks many of the boxes that define healthy aging for us:

- She's stayed interested in her work over many years. Although she's officially past retirement age, she still goes into the office at her research institution.
- She socializes with all sorts of people. She hosts monthly dinner discussions (held in many languages) for her younger colleagues.
- She lives in a fifth-floor walkup apartment. At times, her younger friends have to forgo a dinner party there because they are too stiff or fatigued to get up all those flights of stairs—but Marie-Jeanne navigates them as deftly as she has for many years.
- She is always interested in new experiences, like visiting museum exhibits that come to town.

You might have your own role model, or your own goals for getting older. Here are some others we've heard:

- "When I'm older, I want to be like the actress Judi Dench, especially the way she played M in the Bond movies:

40

white-haired but completely in command, and the smartest person in the room."

■ "I'm inspired by the idea of life's 'third act.' The first act of my life was all about educating myself; the second act was about growing my teaching career; and for my third act I am planning to work with not-for-profits to help teen parents stay in school and complete their degrees."

■ "My grandfather took us kids cross-country skiing well into his seventies and showed us how to build fires in the snow. I want to do the same for my own grandchildren."

■ "As I picture getting older, I imagine that the kids are grown up and out of the house. I miss them, but I've got more time. I can finally accept the offer to chair my department."

■ "If I'm still intellectually curious, and actively working on writing projects or a philanthropic project, I'll be happy. I want to be giving back in more than one way, appreciating our beautiful planet and the best in others, including myself."

Your cells are going to age. But they don't need to age before their time. What most of us really want is to have a long, satisfying life, with advanced cellular aging pushed toward the very end.

The chapter you've just read showed you how prematurely aging cells can hurt you. Next, we're going to show you exactly what telomeres are and how they can give you the best shot at a long, good life.

CHAPTER TWO

The Power of Long Telomeres

It's 1987. Robin Huiras is twelve years old and standing on her school's playing field, waiting to begin a timed mile-long run. The weather is good for running—it's a chilly Minnesota morning—and Robin is fit and slender. Although she doesn't enjoy being put through her paces by the gym teacher, she expects to do well.

She doesn't. The gym teacher fires the starting pistol and almost immediately the other girls in the class are ahead of Robin. She tries to catch them, but the pack recedes along the red-dirt running path. Robin is no slacker—she gives everything she's got, but as the race goes on she falls farther and farther behind. Her final time is one of the slowest in the class, almost as if she'd stopped partway through the course and taken a leisurely stroll across the finish line, but long after the race is finished Robin is still doubled over from the exertion, gulping for air.

The next year, when Robin is thirteen, she spots a gray strand threading its way through her brown hair. Then another gray hair appears, and another, until her hair takes on the light salt-and-pepper appearance that's common among women in their forties or fifties. Her skin changes, too—there are days when normal activities leave deeply colored bruises on her arms and legs. Robin is only a teenager, but her energy is low, her hair is turning gray, and her skin is fragile. It's as if she's growing old before her time.

In a very real way, that was exactly what was happening. Robin has a rare telomere biology disorder, an inherited disorder that causes extremely short telomeres and, in turn, early cell aging. Well before people with telomere biology disorders are chronologically old, they can experience rapidly accelerated aging. Outwardly, it shows up in the skin. Melanocytes, for example, which are the skin's color cells, lose their ability to keep the skin evenly toned. The result is age blotches and spots, along with gray or white hair, even at a young age. Fingernails and toenails look old, too. Because nails have cells that turn over quickly, they become ridged and split. Bones grow older, too: osteoblasts—cells that your bones need to stay solid and strong—can stop renewing themselves. Robin's father, who had the same telomere biology disorder, had so much bone loss and muscle pain that he needed both of his hips replaced twice before the disorder took his life at age forty-three.

But an aged appearance and even bone loss are some of the milder effects of telomere biology disorder. The more devastating ones can include scarred lungs, unusually low blood counts, a weakened immune system, bone-marrow disorders, digestive problems, and certain cancers. People with telomere disorders do not tend to live full life spans, though the precise symptoms and the average length of life varies; one of the oldest known telomere disorder patients alive right now is in her sixties.

Severe inherited forms of telomere biology disorders such as Robin's are an extreme form of much more common conditions, which we now collectively call "telomere syndromes." We understand which genes accidentally go wrong to cause these inherited, severe forms, and what these genes do in cells. (Eleven such genes are known to date.) Thankfully, these extreme, inherited telomere syndromes are rare; they affect about one in a million people. And thankfully, Robin was eventually able to take advantage of medical advances and undergo a successful stem cell transplant (one which contained a donor's blood-forming stem cells). One testimony to the transplant's success is Robin's platelet count. Because Robin's blood stem cells could not effectively repair their telomeres, or make new cells, her platelets had plummeted to alarmingly low numbers, with counts as low as 3,000 or 4,000. (Low blood counts are a reason she couldn't keep up during the mile run.) Six months after the transplant, Robin's counts shot up to more normal levels of almost 200,000. Robin, who is now in her thirties and runs an advocacy organization for people with telomere biology disorders, has more wrinkles around her mouth and eyes than other people her age. Her hair is almost entirely gray, and she sometimes experiences severe joint and muscle pain. But habitual exercise helps keep the pain at bay, and the transplant has restored much of her energy.

Severe inherited telomere syndromes carry a powerful message for all of us, because what is happening inside Robin's cells is also happening inside your own. It's just happening to her faster than it's happening to you. In all of us, telomeres shrink with age. And premature cellular aging can happen—in a slower way—to basically healthy people. We can think of all of us as being susceptible, to some extent, to telomere syndromes of aging, although to much lesser degrees than Robin and her father. Patients with the inherited telomere syndromes are powerless to stop the premature aging process, because it takes place with overwhelming speed in their bodies, but the rest of us are luckier. We have much more control over premature cellular aging, because—to a surprising extent—we have some real control over our telomeres.

That control begins with knowledge—knowledge about telomeres and how their length corresponds to your daily habits and health. To understand the role telomeres play in your body, we need to turn to an unlikely source. We need to spend some time with pond scum.

POND SCUM SENDS A MESSAGE

Tetrahymena is a single-celled organism that swims valiantly through bodies of freshwater, searching for food or a mate. (There are seven sexes of *Tetrahymena*, a curious fact to ponder next time you are splashing around in a lake.) *Tetrahymena* is, literally, pond scum. Yet it's almost adorable. Seen under a microscope, it boasts a plump little body and hairlike projections that make it look like a fuzzy cartoon creature. Look at it long enough, and you might notice a resemblance to Bip Bippadotta, the wild-haired Muppet who scats the famously infectious song "Mahna Mahna."

Inside *Tetrahymena*'s cell is its nucleus, its central command center. Deep within that nucleus is a gift to molecular biologists: twenty thousand tiny chromosomes, all identical, linear, and very short. That gift makes it relatively easy to study *Tetrahymena*'s telomeres, those caps at the ends of chromosomes. That gift is the reason that in 1975, I (Liz) was standing in a laboratory at Yale, cultivating millions of tiny *Tetrahymena* in big glass lab jars. I wanted to collect enough of their telomeres to understand just what they were made of, at the genetic level.

For decades, scientists had theorized that telomeres protect chromosomes—not just in pond scum but in humans, too—but no one knew exactly what telomeres were or how they worked. I thought that if I could pinpoint the structure of the DNA in telomeres, I might be able to learn more about their function. I was driven by my desire to understand biology; at this point, no one knew that telomeres would prove to be one of the primary biological foundations of aging and health.

By using a mixture of what was essentially dish detergent and salts, I was able to release *Tetrahymena*'s DNA from its surrounding matter, out of the cell. Then I analyzed it, using a combination of the chemical and biochemical methods that I'd learned during my PhD graduate years in Cambridge, England. Under the dim, red, and warm safelight of the lab's darkroom, I reached my goal. The darkroom was quiet; only a trickle of water sounded as it ran next to the old-fashioned developing tanks. I held a dripping X-ray film up to the safety light, and excitement surged through me as I understood what I was seeing. At the ends of chromosomes was a simple, repeated DNA sequence. The same sequence, over and over and over. I had discovered the structure of telomere DNA. And, in the ensuing months, as I toiled over pinpointing its details, an unexpected fact rose up: Remarkably, these tiny chromosomes were not as identical as they had seemed. Some had ends with more, and some had ends with fewer numbers of the repeats.

No other DNA behaves in this strangely variable, sequential, repeating way. The telomeres of pond scum were sending a message: There

Figure 7: *Tetrahymena*. This tiny one-celled creature, which Liz studied to decode the DNA structure of telomeres and to discover telomerase, provided the first precious information about telomeres, telomerase, and a cell's life span. This foreshadowed what was later learned in humans.

is something special here at the ends of chromosomes. Something that would turn out to be vital for the health of human cells. That variability in the lengths of the ends turns out to be one of the factors that explains why some of us live longer and healthier than others.

TELOMERES: THE PROTECTORS OF OUR CHROMOSOMES

It became clear from that dripping X-ray film that telomeres are composed of repeated patterns of DNA. Your DNA consists of two parallel, twisting strands that are made up of just four building blocks ("nucleotides") that are represented by the letters A, T, C, and G. Remember grade school field trips, when you had to hold hands with a buddy as you walked through a museum? The letters of DNA operate on the buddy system, too. A always pairs with T, and

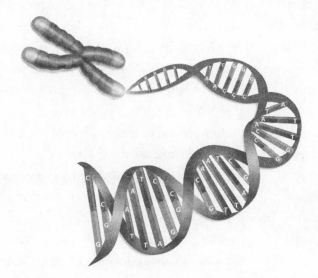

Figure 8: Telomere Strands Up Close. At the tips of the chromosome are the telomeres. The telomere strand is made up of repeating sequences of TTAGGG that sit across from their base pair partners, AATCCC. The more of these sequences we have, the longer our telomeres. In this diagram we depict just the DNA of telomeres, but it is not bare like this—it is covered by a protective sheath of proteins.

C always pairs with G. The letters on the first strand of DNA pair up with their partners on the second strand. The two make up a "base pair," which is the unit we measure telomeres in.

In human telomeres (as would later be discovered), the first strand consists of repeating sequences of TTAGGG, and they are coupled with their pairs, AATCCC, on the second strand, twisted into the helix shape that is DNA.

These are the base pairs of telomeres that, repeated thousands of times, offer a way of measuring their length. (Note some of our graphs measure telomere length in a unit called a t/s ratio, instead of base pairs, which is just another way to measure telomeres.) The repeating sequence highlights the differences between telomeres and other DNA. Genes, which are made of DNA, live within a chromosome. (Inside a cell we have twenty-three pairs of chromosomes, for a total of forty-six.) This genetic DNA is what forms your body's blueprint, its instruction manual. Its paired letters create complicated "sentences" that send instructions for building the proteins that make up your body. Genetic DNA can help determine how quickly your heart beats, whether your eyes are brown or blue, and whether you're going to have the long legs and arms of a distance runner. The DNA of telomeres is different. First of all, it doesn't live inside any gene. It sits outside of all the genes, at the very edges of the chromosome that contains genes. And unlike genetic DNA, it doesn't act like a blueprint or code. It's more like a physical buffer; it protects the chromosome during the process of cell division. Like beefy football players who surround a quarterback, absorbing the hardest blows from the onrush of opposing players, telomeres take one for the team.

This protection is crucial. As cells divide and renew, they need their precious chromosome cargoes of genetic instruction manuals (the genes) to be delivered intact. Otherwise, how would a child's body know to grow tall and strong? How would your cells know to produce the body traits that make you feel like *you*? Yet cell division is a potentially dangerous time for chromosomes and the genetic material inside. Without protection, chromosomes and the genetic material they carry could easily become unraveled. The chromosomes can break, can fuse with others, or can mutate. If your cell's genetic instruction manuals were scrambled like this, the result would be disastrous. A mutation can lead to cell dysfunction, cell death, or even proliferation of a now-cancerous cell, and as a consequence you probably wouldn't live very long.

Telomeres, which seal off the ends of the chromosomes, keep this unthinkable event from happening. That is the message sent to us by the special repeating sequences of telomere DNA. Jack Szostak and I (Liz) discovered this function in the early 1980s, when I isolated a telomere sequence from *Tetrahymena* and Jack put it into a yeast cell. The *Tetrahymena* telomeres protected the chromosomes of the yeast during cell division by donating some of their own base pairs.

Every time a cell divides, its precious "coding DNA" (which makes up the genes) is copied so it can stay safe and whole. Unfortunately, with each division, telomeres lose base pairs from the sequences at the two ends of each chromosome. Telomeres tend to shorten as we get older, as our cells experience more and more divisions. But the trend is not just a straight line. Take a look at the graph on the next page.

In the Kaiser Permanente Research Program on Genes, Environment, and Health study of one hundred thousand people's salivary telomere length, telomeres on average grew shorter and shorter as people progressed from their twenties, hitting rock bottom at around age seventy-five.[1] In an interesting coda, telomere length appears to stay the same or even go *up* as people live past seventy-five. This trend is probably not true lengthening happening; it just looks that way because the folks with shorter telomeres have passed away by this age (which is called survival bias—in any aging study, the oldest people are the healthy survivors). It's the people with longer telomeres who are living into their eighties and nineties.

49

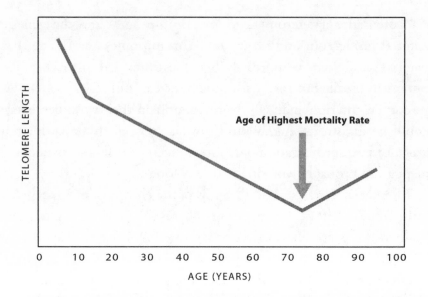

Figure 9: Telomeres Shorten with Age. Telomere length declines with age, on average. It declines fastest during early childhood and then has a slower average rate of decline with age. Interestingly, many studies find telomere length is not shorter in those who live to be a lot older than seventy years. This is thought to be due to "survival bias," meaning that those still alive at this age tended to have been those people with longer telomeres. Their telomeres probably had been longer all along, starting from birth.

TELOMERES, THE DISEASESPAN, AND DEATH

Telomeres shorten with age. But can our telomeres really help determine how long we'll live or how soon we enter the diseasespan?

Science says *yes*.

Short telomeres don't predict death in every study, since there are many other factors that predict when we die. They do predict time of death in around half of studies, including the largest study yet. A 2015 Copenhagen study of more than sixty-four thousand people shows that short telomeres predict earlier mortality.[2] The shorter your telomeres, the higher your risk of dying from cancer, cardiovascular disease, and of dying at younger ages generally, known as all-cause mortality. Look at figure 10, and you'll see that telomere length is broken out by percentiles in ten groups. People in the 90th percentile of telomere length (with the longest telomeres) are at the left; people in the 80th percentile are just next to them; and so on all the way to the right side, where the people in the lowest percentile are represented. There is a graded response: people with the longest telomeres are the healthiest, and as telomeres get shorter, people get sicker and are more likely to die.

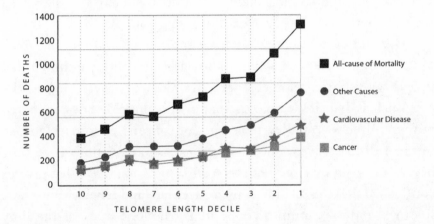

Figure 10: Telomeres and Death. Telomere length predicts mortality overall, and from different diseases. Those with the longest telomeres (90th percentile) have the lowest rate of death from cancers, heart disease, and all causes added up. (Figure is from the data in Rode et al., 2015.[3])

The Kaiser Permanente study previously mentioned measured telomere length in one hundred thousand research volunteers who happened to be members of Kaiser's health coverage plan. In the three years after their telomeres were measured, the people with shorter telomeres were more likely to die when all causes of death were combined.[4] The study was controlled for differences among the subjects that might likely lead to differences in health and longevity, including age, gender, race and ethnicity, education, smoking, physical activity, drinking, and body mass index (BMI). Why did the scientists control for so many variables? Because any one, some, or all of these factors might in theory have been the real reasons contributing to the increased mortality, not the shortened telomeres. For example, a clear relationship exists between tobacco smoking histories and all-cause mortality rates. And many studies have found a relationship between more smoking and more telomere shortening. Yet even after correcting for all those potential explanations, the relationship between telomere shortness and all-cause mortality still held true. It does indeed look as though telomere shortness itself is a real contributor to our overall risks for mortality.

Over and over and over, telomere shortness has also been linked to the major diseases of aging. Many large studies have shown that people with shorter telomeres are more likely to have a chronic disease, such as diabetes, cardiovascular disease, lung diseases, impaired immune function, and certain types of cancers, or to develop one of these diseases over time.[5] Many of these associations have now been reinforced by large reviews (called meta-analyses) that give us confidence that the relationships are accurate and reliable. Flip these findings, and the optimistic opposite is true: one study of a healthy elderly U.S. sample (the Health ABC study) showed that in the general population, people with longer telomeres in their white blood cells had more years of healthy life without any major diseases—a longer healthspan.[6]

TURN THE TIDE IN HEALTH

People like Robin Huiras, whose rare inherited disorder leads to telomeres that are drastically short, show us the power of telomeres. Sometimes, as in Robin's case, it's a kind of dark, corroding power that speeds up the cellular aging process. The good news is that we have learned a great deal about the nature of telomeres. By donating blood and tissue samples, for example, Robin and her family have helped researchers pinpoint one of the gene mutations that caused her disorder. That knowledge is a first step to better diagnoses, treatments, and, one day, a cure.

And you can use our knowledge about telomeres to turn the tide in health—in your health, the health of people in your community, and the health of generations to come. Because as you're about to see, telomeres can change. *You* have the power to influence whether your telomeres are going to shorten early, or whether they are going to stay supported and healthy. To show you what we mean, we need to take you back to Liz's lab. There, *Tetrahymena* telomeres began to behave in a strange, unexpected way.

CHAPTER THREE

Telomerase, the Enzyme That Replenishes Telomeres

Not long after I (Liz) had read the X-ray revealing the DNA of telomeres, I was hired by the University of California, Berkeley, where in 1978 I set up my own laboratory to continue my research into telomeres. There I began to notice something that shocked me. I was still growing *Tetrahymena,* that hairy, Muppet-like pond scum, and now I was able to tell the sizes of their telomeres from the length of their DNA. And mysteriously, under some conditions, the *Tetrahymena*'s telomeres would sometimes *grow.*

This was a shock because I expected that if telomeres were going to change at all, they were going to get shorter, not longer: with each cell division, the number of DNA sequences in the telomeres would more likely shrink. Yet it looked to me as if *Tetrahymena* was creating new DNA. But this was not supposed to happen. DNA is not supposed to change. You've probably heard that the DNA we are born with is the DNA we die with, and that DNA is produced solely through a kind of biochemical photocopying. I checked, and double-checked, and confirmed that what was supposed to be impossible was, in fact, happening. Next, we saw the same thing happening in yeast cells, too. ("We" here included my student Janice Shampay in my lab, working on the experiments that Harvard researcher Jack Szostak and I had dreamed up together.) Then reports from other scientists trickled in that suggested these changes might happen in other tiny, *Tetrahymena*-like creatures, too. The organisms were, in fact, producing new DNA at the ends of their telomeres. Their telomeres were growing.

No other element of DNA behaves in this way. For decades, genetic scientists believed that any stretch of chromosomal DNA existed only because it had been copied from preexisting DNA. The accepted wisdom was that DNA could not be created from whole cloth where there had not been DNA before. The discovery of this odd behavior told me *there was something going on here that no one had seen before.* For scientists, that's one of the most exciting kinds of discoveries to make. It's thrilling when a strange finding suggests that there are new, unknown street corners of the universe, ripe for exploration. As it turned out, this behavior of telomeres led to more than just a new street corner of the universe; this was a whole new neighborhood, one that no one had known existed.

TELOMERASE: THE SOLUTION TO TELOMERE SHRINKAGE

I kept pondering this strange behavior of the telomere, its apparent ability to grow. I wanted to look for an enzyme in a cell that might add DNA onto telomeres—an enzyme that might replenish telomeres after they'd lost some of their pairs of letters. It was time for me to roll up my sleeves and make more *Tetrahymena* cell extracts. Why *Tetrahymena*? Because it's such a good source of plentiful telomeres. I reasoned that it might be a good source of enzymes that could form telomeres, if such an enzyme existed.

In 1983 I was joined in this quest by Carol Greider, a new graduate student in my lab. We began devising experiments, and then refining those experiments, and on Christmas Day in 1984, Carol developed an X-ray film called an autoradiograph. The patterns on that film showed the first clear signs of a new enzyme at work. Carol went back home and danced with excitement in her living room.

The next day, her face alight with suppressed glee in her anticipation of my reaction, she showed me the X-ray film. We looked at each other. Each of us knew this was it. Telomeres could add DNA by attracting this previously undiscovered enzyme, which our lab named telomerase. Telomerase creates new telomeres patterned on its own biochemical sequence.

But science does not work only by the exhilaration of one single eureka moment. We had to be sure. As the weeks stretched into months, we experienced surges of doubt followed by thrills of joy as we painstakingly conducted follow-up experiments. Step by step, we ruled out every possible reason that our exciting first moments in 1984 could have been just a false lead. Eventually, a deeper understanding of telomerase emerged: Telomerase is the enzyme responsible for restoring the DNA lost during cell divisions. Telomerase makes and replenishes telomeres.

Here's how telomerase works. It includes both protein and RNA, which you can think of as a copy of DNA. That copy includes a template of the telomere's DNA sequence. Telomerase uses that sequence in the RNA as its own inbuilt biochemical guide to create the right sequence of brand-new DNA. The right sequence is needed to make a DNA scaffold perfectly shaped to attract a sheath of telomere-protective proteins that cover the telomeric DNA. This new segment of DNA is added by telomerase to the end of the chromosome, guided by the RNA template sequence and the DNA's buddy partner system of pairing up its letters. This ensures that the right sequence of building blocks of telomeric DNA is added. In this way, telomerase re-creates new endings at the chromosome's tips and replaces ones that have been worn down.

The mystery of the growing telomeres was solved. Telomerase replenishes telomeres by adding telomeric DNA to them. Each time a cell divides, telomeres gradually shorten until they reach a crisis point that signals cell division to stop. But telomerase counteracts this telomere shortening by adding DNA and building back the chromosome end each time a cell divides. This means that the chromosome itself is protected, and an accurate copy of it is made for the new cell. The cell can continue to renew itself. **Telomerase can slow, prevent, or even reverse the shortening of telomeres that comes with cell division.** Telomeres can, in a sense, be renewed by telomerase. We had found a way to get around the Hayflick limit of cell division... in pond scum.

TELOMERASE: NO ELIXIR OF IMMORTALITY

After these discoveries, both the scientific world and the global media buzzed with hopeful speculation. What if we could increase our supply of telomerase? Could we be like *Tetrahymena*, with cells that renew forever? (This may have been the first recorded instance of humans fervently wishing to be more like pond scum.)

People wondered if telomerase could be distilled and served up as an elixir of immortality. In this wishful scenario, we'd visit our local telomerase bar every now and then for a hit of the enzyme, which would let us live healthy lives all the way to the very end of the known maximum human life span—or beyond it.

These dreams are perhaps not as ridiculous as they might seem. Telomeres and telomerase form a crucial biological foundation for cell aging. The demonstration of the relationship between telomerase and cell aging first came from *Tetrahymena*. Guo–Liang Yu, then a graduate student in my Berkeley lab, performed a simple but surgically precise experiment. He replaced the normal telomerase in *Tetrahymena* cells with a precisely inactivated version. If you feed them properly, *Tetrahymena* cells are normally immortal in the laboratory. Like the Energizer Bunny, *Tetrahymena*'s cell divisions normally just keep going and going and going. But this inactivated telomerase caused the telomeres to become shorter and shorter as the *Tetrahymena* cells divided. Then when the telomeres had become too short to protect the genes inside the chromosome, the cells stopped dividing. Think again of a shoelace. It's as though the shoelace tip wore down and the shoelace—with all that vital genetic material—became frayed. Inactivating telomerase made the *Tetrahymena* cells mortal.

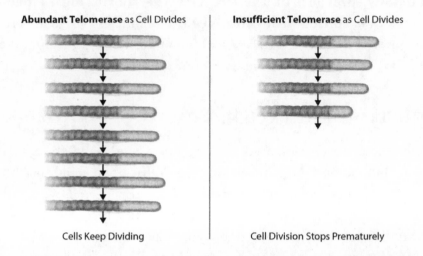

Abundant Telomerase as Cell Divides | **Insufficient Telomerase** as Cell Divides

Cells Keep Dividing | Cell Division Stops Prematurely

Figure 11: Consequences of Enough, or Not Enough, Telomerase Action. Telomere DNA shortens because the enzymes to duplicate the DNA don't work at the telomere ends (incomplete DNA replication). Telomerase elongates telomeres and thus counterbalances the inexorable attrition of telomeric DNA. With abundant telomerase, telomeres are maintained and cells can keep dividing. With insufficient telomerase (due to genetics, lifestyle, or other causes) telomeres shorten rapidly, cells stop dividing, and senescence soon follows. Reprinted with permission from AAAS (Blackburn, E., E. Epel, and J. Lin., "Human Telomere Biology: A Contributory and Interactive Factor in Aging, Disease Risks, and Protection," *Science* (New York) 350, no. 6265 (December 4, 2015): 1193–98).

Without telomerase, the cells stop renewing.

And then, at other labs around the world, the same was found for nearly all cells, except bacteria (whose chromosomes are circles of DNA instead of lines and thus have no ends to protect). Longer telomeres and more telomerase delayed premature cellular aging, and shortened telomeres and less telomerase sped it up. The telomerase-health connection was nailed when clinician Inderjeet Dokal and his colleagues in the United Kingdom and the United

States first discovered that when people have a genetic mutation that slashes telomerase levels in half, they develop severe inherited telomere syndromes.[1] This is the same category of disease that was diagnosed in Robin Huiras. Without sufficient telomerase, the telomeres quickly shorten, and the body succumbs to early disease.

Tetrahymena cells have telomerase in sufficient quantities so that they can constantly rebuild their telomeres. This allows *Tetrahymena* to perpetually renew itself and to forever avoid cell aging. But we humans normally don't have enough telomerase to accomplish this feat. We are very miserly when it comes to telomerase. Our cells are reluctant to hand out telomerase willy-nilly to their telomeres all the time. We produce telomerase in sufficient quantities to rebuild telomeres... but only up to a point. As we age, the telomerase in most of our cells generally becomes less active, and telomeres get shorter.

TELOMERASE AND THE CANCER PARADOX

It's natural to wonder if we could extend human life through artificial methods of increasing telomerase. Ads for telomerase-boosting supplements abound on the Internet claiming that we can. Telomerase and telomeres have wonderful properties that can allow us to avoid horrible diseases and feel more youthful. But they are not magical life extenders—they don't let us live past the normal human life span as we know it. In fact, if you try to extend your life by using artificial methods of increasing telomerase, you are putting yourself in danger.

That's because telomerase has a dark side. Think of Dr. Jekyll and Mr. Hyde—they are the same person, but one with a drastically different character depending on whether it is day or night. We need our good Dr. Jekyll telomerase to stay healthy, but if you get too much of it in the wrong cells at the wrong time, telomerase takes on its Mr. Hyde persona to fuel the kind of uncontrolled cell growth that is a hallmark of cancer. Cancer is, basically, cells that won't stop dividing; it's often defined as "cell renewal run amok."

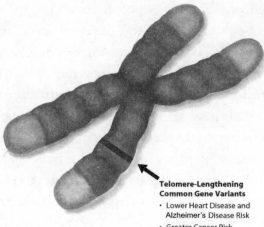

Telomere-Lengthening
Common Gene Variants
• Lower Heart Disease and
  Alzheimer's Disease Risk
• Greater Cancer Risk

Figure 12: Telomere-Related Genes and Disease. Telomere maintenance genes can protect us from common diseases, but can put us at risk for some cancers. Having gene variants for more telomerase and telomere proteins means longer telomeres. This natural genetic way of making telomeres longer lowers risks for most diseases of aging, including heart disease and Alzheimer's disease, but the high telomerase also means that cells that are prone to become cancerous can keep dividing unchecked, causing a greater risk for certain types of cancer (brain cancers, melanoma, and lung cancers). Bigger isn't always better!

You don't want to bomb your cells with artificial telomerase that may goad them into taking the road toward becoming cancerous. Unless the telomerase supplement field comes up with more thorough demonstrations of safety in large—and *long-term*—clinical trials, in our view it's sensible to skip any pill, cream, or injection that claims it will increase your telomerase. Depending on your individual propensity for different types of cancer, you may be potentially increasing the chance of developing any of a number of different cancers (such as melanomas or brain and lung cancers). Knowing this, it comes as no surprise that our cells keep their telomerase on a tight rein.

Given these scary-sounding findings, you may be wondering, why are we suggesting activities that boost telomerase? The answer is that there is a big difference between the body's normal physio-logical responses to the lifestyle suggestions we make for your health in this book and taking an artificial substance (no matter how "nat-ural" its plant source—remember that plants are some of nature's biggest chemical warfarers, having evolved an armamentarium of strong chemicals to fend off hungry animals and marauding patho-gens). The suggestions we include in this book for increasing your telomerase action are gentle and natural—and they increase telo-merase in safe amounts. You do not need to worry about an increased cancer risk with these strategies. They simply don't increase telome-rase to the levels or in the ways that would be harmful.

Paradoxically, we do need to keep our telomeres healthy to ward off cancer, too. Some types of cancers are more likely to develop when *too little available telomerase* makes telomeres too short—blood cancers like leukemia, skin cancers besides melanoma, and some gas-trointestinal cancers such as pancreatic cancer. This was proved with the discovery that people born with a mutation that precisely inacti-vated a telomerase gene had much higher risk for these cancers. Such cancers arise because losing telomere protection allows our genes to become more easily damaged—and altered genes can eventually lead to cancers. Furthermore, too little telomerase weakens the telomeres in our immune cells. Our immune system usually keeps a sharp eye out for anything perceived as "foreign," and that includes harmful cancer cells as well as pathogenic invaders from the outside such as bacteria and viruses. Without telomeres long enough to act as buffers, the cells of the immune system will eventually become senescent.

Some of these immune cells are like surveillance cameras posted at every corner of the body. If they become senescent, their lenses then act as though they are steamed up, and they miss the "foreign" cancer cells. So the teams of immune cells that would normally be called up fail to leap into action. The result of weakened telomeres is that the immune defenses of the body are more likely to lose the fight against a cancer (or a pathogen).

TELOMERASE AND HOPE FOR
NEW CANCER TREATMENTS

Too much telomerase, spurred by the actions of even normal variants of telomerase genes, can increase risks of developing several forms of cancer. And overactive telomerase fuels most cancers once they turn malignant. But even this "dark side" of telomerase may not always be dark. Researchers have learned that telomerase is hyperactive in roughly 80 to 90 percent of malignant human cancers, with levels that are turned up ten to hundreds of times as high as in normal cells. This discovery may one day turn out to be a potent weapon in our fight against the disease. If telomerase is necessary for cancers to grow so relentlessly, perhaps we can treat cancer by turning off the telomerase in just the cancer cells. Researchers are at work on this idea.

The key is to well regulate the action of telomerase on telomeres—in the right cells and at the right times, because only that will keep telomeres and us healthy. **The body knows how to do this, and we can help it with a lifestyle full of renewal strategies.**

YOU CAN INFLUENCE YOUR TELOMERES AND TELOMERASE

By the turn of the millennium, scientists had become accustomed to thinking about both telomeres and telomerase as foundations of cell renewal. But the telomere syndromes, starting with the shocking finding that cutting telomerase by merely half could have such drastic effects, had galvanized everyone into thinking only in terms of *genes* that determined whether our telomeres were long or short, and whether we had enough telomerase to replenish worn-down telomeres.

That was when I (Elissa) began a postdoctoral fellowship in

health psychology at the University of California, San Francisco. Susan Folkman, the now-retired director of the Osher Center for Integrative Medicine and a pioneer in the study of stress and coping, invited me to join a team that was interviewing mothers of children with chronic conditions, a group under tremendous psychological strain.

I felt a profound empathy for these caregiving mothers, who seemed extraordinarily worn out and older than their chronological ages. By then, Liz had moved to the San Francisco campus of the University of California, and I was aware of her work on biological aging. I approached Liz and told her about the caregiving mothers we were studying. If I could come up with the funds, would it be possible to test the mothers' telomeres and telomerase? Was it worth investigating whether stress could shorten telomeres and lead to early cell aging?

Like most other molecular biologists at the time, I (Liz) was peering down at telomeres from one particular mountaintop. I was thinking about our telomere maintenance in terms of the cellular molecules specified by the genes that control telomeres. When Elissa asked me about studying caregivers, however, it was as though I suddenly saw telomeres from a whole new viewpoint, from a completely different mountain. I responded as both a scientist and a mother. "We need another ten years just to more fully understand the genetics of telomeres," I mused, somewhat doubtfully, but I could also well imagine the tremendous stresses these women were under. I thought about the way we describe exhausted, stressed people: *careworn*. Mothers of chronically sick children are women who are worn down. Was it possible that their telomeres were worn down, too? "Yes," I agreed. "Let's do this study, if we can find a scientist in my lab who will help with the measurements." My postdoctoral fellow, Jue Lin, raised her hand. She proceeded to refine a way to sensitively and carefully measure telomerase in healthy human cells, and the work began.

We selected a group of mothers who were each caring for a chronically ill biological child. A research subject who might have an outside "issue" could warp the results, so any mother with a major health problem was screened out of the study. We used a similar process to select a control group of mothers whose children were healthy. This process took several years of careful selection and assessments.

We took a sample of each woman's blood and measured the telomeres in the white blood cells. We recruited the help of Richard Cawthon at the University of Utah, who had recently devised a new easier way to measure the length of telomeres in white blood cells (applying a method called polymerase chain reaction).

One day in 2004, the assay results came in. I (Elissa) was sitting in my office as the numerical analysis came out of the printer. I looked down at the scatter plot and gasped. There was a pattern to the data, the exact gradient we thought might exist indeed was right there, on the page. It showed that the more stress you are under, the shorter your telomeres and the lower your telomerase levels.

I immediately picked up the phone and called Liz. "The results are in," I said, "and the findings are even more striking than we'd thought they might be."

We'd asked the question *Can the way we live change our telomeres and telomerase?* Now we had an answer.

Yes.

Yes, the mothers who perceived themselves to be under the most stress were the ones with the lowest telomerase.

Yes, the mothers who perceived themselves to be under the most stress were the ones with the shortest telomeres.

Yes, mothers who had been caregiving for the longest time had shorter telomeres.

This triple yes meant that our results weren't just a coincidence or a statistical blip. **It also meant that our life experiences, and the way we respond to those events, can change the lengths**

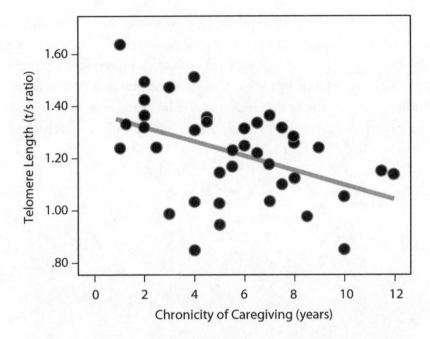

Figure 13: Telomere Length and Chronic Stress. The more years since the child had been diagnosed (thus the more years of chronic stress), the shorter the telomeres.[2]

of our telomeres. In other words, we can change the way that we age, at the most elemental, cellular level.

Whether aging can be sped up, slowed, or reversed has been a topic of medical debate for centuries. What we have learned since this first study of caregivers is wholly new. We, as a field, have learned that by our actions, we can keep our telomeres—and hence our *cells*—from aging prematurely. We may even be able to partly reverse the cellular aging process caused by telomere wear and tear. Over the years, the results of our initial study have held up, and many additional studies that you will read about here have taken this first finding much further, showing that many different life factors can affect our telomeres.

In the rest of this book you will hear us talking about how you can increase telomerase and protect your telomeres. Our recommendations are based on studies, some that measure telomeres, some that measure telomerase activity, and some that do both. You can join us on the journeys of exploration that have followed from those first mountaintop views. Use this research as a North Star to help you change the way you use your mind, take care of your body, and even how you interact with your community, to protect your telomeres and enjoy your healthspan.

RENEWAL LABS: A Guide

Life is full of little experiments we can learn from. Throughout the rest of this book, you will find a Renewal Lab at the end of each chapter. There, you become the researcher if you wish. Your mind, your body, and your life become your personal laboratory where you can try out practical applications of telomere science or behavioral science and learn ways to change your daily life to enhance your cellular health. In most cases, the Renewal Labs have been directly linked to better telomere length, and in all cases they are associated with better physical or mental health. (You can find the relevant studies in the book's Notes section, which begins on page 337.)

When we say "laboratory," we really mean it. They're experiments, not written-in-stone commandments. What works best for you depends on your individual mind and body, your preferences, and your stage of life. So give them a try—perhaps only one or two at a time. If you find one that works for you, focus on it for a while, until it becomes a habit. If you practice any of these Labs regularly, they should enhance your cellular health as well as your daily wellbeing. Studies have found that lifestyle changes can have an effect on telomere maintenance (that means increased telomerase or telomere length) as soon as three weeks to four months. Remember, as Ralph Waldo Emerson said, "Don't be too timid or squeamish about your actions. All life is an experiment. The more experiments you make the better."

PART II

YOUR CELLS ARE LISTENING TO YOUR THOUGHTS

ASSESSMENT: Your Stress Response Style Revealed

Part Two, "Your Cells Are Listening to Your Thoughts," offers insights into how you experience stress and how you can shift that experience to be healthier for your telomeres and more beneficial in your daily life. To get you started, here's a quick self-test. It assesses your underlying sources of stress reactivity and stress resilience, some of which have been linked to telomere length.

Think of a situation that bothers you a great deal and that is ongoing in your life. (If you cannot think of a current situation, think of your most recent difficult problem.) Circle your numerical response to each question.					
1. When you think about dealing with this situation, how much do you feel hope and confidence vs. feelings of fear and anxiety?	**0** hopeful, confident	**1**	**2** same amount of each	**3**	**4** fearful, anxious
2. Do you feel you have whatever it takes to cope effectively with this situation?	**4** not at all	**3**	**2** somewhat	**1**	**0** extremely
3. How much are you caught up in repetitive thoughts about this situation?	**0** not at all	**1**	**2** somewhat	**3**	**4** extremely
4. How much do you avoid thinking about the situation or try not to express negative emotions?	**0** not at all	**1**	**2** somewhat	**3**	**4** extremely
5. How much does this situation make you feel bad about yourself?	**0** not at all	**1**	**2** somewhat	**3**	**4** extremely
6. How much do you think about this situation in a positive way, seeing some good that could come from it, or telling yourself statements that feel comforting or helpful, such as that you are doing the best you can?	**4** not at all	**3**	**2** somewhat	**1**	**0** extremely
TOTAL SCORE (Add up the numbers; notice questions 2 and 6 are positive responses so the scale is reversed.)					

The point of this informal test (not a validated research measure) is to raise awareness of your own tendencies to respond in a certain way to chronic stress. It is not a diagnostic scale. Also know that if you're dealing with a severe situation, your response style score will naturally shift to be higher. This is not a pure measure of response style, because our situations and our responses inevitably get a bit mixed together.

Total score of 11 or under: Your stress style tends to be healthy. Instead of feeling threatened by stress, you tend to feel challenged by it, and you limit the degree to which the situation spills over into the rest of your life. You recover quickly after an event. This stress resilience is positive news for your telomeres.

Total score of 12 or over: You're like most of us. When you're in a stressful situation, the power of that threat is magnified by your own habits of thinking. Those habits are linked, either directly or indirectly, to shorter telomeres. We'll show you how to change those habits or soften their effects.

★ ★ ★

Here's a closer look at the habits of mind associated with each question:

Questions 1 and 2: These questions gauge how threatened you feel by stress. High fear combined with low coping resources turn on a strong hormonal and inflammatory stress response. **Threat stress** involves a set of mental and physiological responses that can, over time, endanger your telomeres. Fortunately, there are ways to convert threat stress into a feeling of challenge, which is healthier and more productive.

Question 3: This item assesses your level of **rumination**. Rumination is a loop of repetitive, unproductive thoughts about something that's bothering you. If you're not sure how often you ruminate, now you can start to notice. Most stress triggers are short-lived, but we humans have the remarkable ability to give them a vivid and extended life in the mind, letting them fill our headspace long after the event has passed. Rumination, also known as brooding, can slip into a more serious state known as depressive rumination, which includes negative thoughts about oneself and one's future. Those thoughts can be toxic.

Question 4: This one's about **avoidance and emotion suppression**. Do you avoid thinking about the stressful situation or avoid sharing feelings about it? Is it so emotionally loaded that the thought of it makes your stomach clench? It's natural to try to push difficult feelings away, but although this strategy may work in the short term, it doesn't tend to help when the situation is chronic.

Question 5: This question addresses "**ego threat**." Does it feel as if your pride and personal identity could be damaged if the stressful situation doesn't go well? Does the stress trigger negative thoughts about yourself, even to the extent that you feel worthless? It's normal to have these self-critical thoughts sometimes, but when they are frequent, they throw the body into an overly sensitive, reactive state characterized by high levels of the stress hormone cortisol.

Question 6: This question asks whether you're able to engage in **positive reappraisal**, which is the ability to rethink stressful situations in a positive light. Positive reappraisal lets you take a less than ideal situation and turn it to your benefit or at least take the sting out of it. This question also measures whether you tend to offer yourself some healthy **self-compassion**.

If the assessment revealed that you struggle with your stress responses, take heart. It's not always possible to change your automatic response, but most of us can learn to change our responses *to our responses*—and that's the secret sauce of **stress resilience**. Now let's get to work understanding how stress affects your telomeres and cells, and how you can make changes that will help protect them.

CHAPTER FOUR

Unraveling: How Stress Gets into Your Cells

We explore the stress–telomere connection, explain toxic stress versus typical stress, and show how stress and short telomeres affect the immune system. People who respond to stress by feeling overly threatened have shorter telomeres than people who face stress with a rousing sense of challenge. Here, you'll learn how to move from harmful stress responses to helpful ones.

Nearly fifteen years ago, my husband and I (Elissa) were driving across the country. We had just finished graduate school at Yale and were taking on postdoctoral fellowships in the Bay Area. San Francisco is an expensive city, and so we had arranged to live with my sister and her family. We expected that when we arrived in San Francisco, we would meet our new nephew, who was supposed to be born at any moment. In fact, he was quite overdue. I called every day for news, but I'd had trouble reaching anyone in the family for days.

About halfway through the trip, just after we'd passed Wall Drug Store in South Dakota, my cell phone finally rang. Tearful voices wavered on the other end. The baby had been born, but something had gone terribly wrong during an induced delivery. The baby was on life support and being fed through a gastric tube to his stomach.

He was a beautiful healthy boy, but an MRI showed his brain had been profoundly damaged. He was paralyzed, blind, and wracked with seizures.

Eventually, after several months, the baby left the intensive care ward and came home. We joined the family team to help take care of this little guy, who had extraordinary needs. We became intimate with both the demands and sorrows that come with a life of caregiving. We were accustomed to pressure and hard work, but this had nothing in common with the types of stresses we had known. Now there were new feelings of constant vigilance, intermittent urgency, worry about the future, and most of all, a heavy weight on the heart. One of the hardest parts was seeing and feeling the pain my sister and brother-in-law were experiencing every day. On top of the emotional suffering there was, all of a sudden, a new, unexpected, and demanding life centered around medical caregiving.

Caregiving is one of the most profound stresses a person can experience. Its tasks are emotionally and physically demanding, and one reason caregivers get so worn out is that they don't get to go home from their caregiving "jobs" and recover. At night, when we all need to biologically check out and refresh body and mind, caregivers are on call. They may be repeatedly woken from sleep to respond to someone in need. Caregivers rarely have time to take care of themselves. They skip their own doctor appointments as well as opportunities for exercise and going out with friends. Caregiving is an honorable role that is taken on out of love, loyalty, and responsibility, but it is not supported by society or recognized for its value. In the United States alone, family caregivers perform an estimated $375 billion in unpaid services each year.[1]

Caregivers often feel unappreciated and become isolated. Health researchers have identified them as one of the most chronically stressed groups of people. This is why we often ask caregivers to volunteer for our studies on stress. Their experiences can tell us a lot about how telomeres react to serious stress. In this chapter, you'll learn what our groups of caregivers have taught us—that chronic, long-lasting stress can erode telomeres. Fortunately for all of us who cannot escape chronic stress (and for all of us who scored higher than 12 on the stress assessment on page 71), we've also learned that we can protect our telomeres from some of stress's worst damage.

"LIKE THERE IS AN ASSAILANT, WAITING FOR ME": HOW STRESS HURTS YOUR CELLS

In our very first study together we looked at some of the most highly stressed caregivers of all: mothers who were taking care of their chronically ill children. This is the study we've told you about. It's the one that first revealed a relationship between stress and shorter telomeres. Now we want to show you a close-up look at the extent of that damage. More than ten years later we still find it sobering.

We learned that the years of caregiving had a profound effect, grinding down the women's telomeres. The longer a mother had been looking after her sick child, the shorter her telomeres. This held true even after we took into account other factors that might affect telomeres, like the mother's age and body mass index (BMI), which are related to shorter telomeres themselves.

There was more. The more stressed out the mothers felt, the shorter their telomeres. This was true not just for the caregivers of sick children, but also for *everyone* in the study, including the control group of mothers who had healthy children at home. The high-stress mothers also had almost half the levels of telomerase than the low-stress mothers, so their capacity to protect their telomeres was lower.

People experience stress in many different ways: "like a fifty-pound weight on my chest," "like a knot in my stomach," "like a vacuum in my lungs that doesn't let me take a full breath," "my heart pounds like there is an assailant, waiting for me." These metaphors are grounded in the body, because stress is as present in the body as in the head. When the stress-response system is on high alert, the body produces more of the stress hormones cortisol and epinephrine. The heart beats faster and blood pressure increases. The vagus nerve, which helps modulate the physiological reaction to stress, withdraws its activity. That's why it's harder to breathe, harder to stay in control, harder to imagine that the world is a safe place. When you suffer from chronic stress, these responses are on a low but constant alert, keeping you in a state of physiological vigilance.

In our caregivers, several aspects of the physiological stress response, including lower vagus activity during stress, and higher stress hormones while sleeping, were linked to shorter telomeres or to less telomerase.[2] These responses to stress appeared to be accelerating the biological aging process. We had discovered a new reason that stressed-out people look haggard and get sick: their heavy stresses and cares are wearing down their telomeres.

SHORT TELOMERES AND STRESS: CAUSE OR EFFECT?

When a scientific finding suggests a cause-and-effect relationship, you have to ask whether the relationship really runs in the direction you think it does. For example, people used to think that fevers caused sickness. Now we know that the relationship is the reverse: sickness causes fevers.

As the results of our first study of caregivers came in, we were careful to ask ourselves *why* shorter telomeres appeared in people with higher stress. Does stress really lead to short telomeres? Or can short telomeres somehow predispose a person to feeling more stress? Our caregiving mothers provided the first convincing data about this question. The relationship between the years of caregiving stress and telomere length is a strong indicator that the stress exposure happens over time, causing telomeres to shorten.

Short telomere length (after correcting for age) could not have determined how many years a mother had been a caregiver, so it had to be the other way around—that the years of caregiving were the cause of the shorter telomeres. We also tested whether an older age of the child was related to shorter telomeres. If the years of difficult caregiving were wearing down telomeres more than the years of parenting by the control mothers, we would see the relationship between the child's age and the mother's telomeres in the caregivers but not in the control moms. Indeed, this was what we found. Now there are animal studies showing that inducing stress can actually cause telomere shortening.

The depression story is more complicated. The findings above were not enough to rule out the possibility that cell aging could cause depression. In humans, depression runs in families. Not only are girls whose mothers have depression more prone to depression themselves, but even before any depression has developed, these girls have shorter blood telomeres than girls who are not depressed.[3] Also, the more stress reactive the girls are, the shorter the telomeres. So the arrow likely points in both directions with depression—short telomeres may precede depression, and depression may speed up telomere shortening.

HOW MUCH STRESS IS TOO MUCH?

Stress is unavoidable. How much of it can we handle before our telomeres are damaged? A consistent lesson from the past decade of studies—and a lesson that echoes what the caregivers taught us—is that stress and telomeres have a dose-response relationship. If you drink alcohol, you're familiar with dose and response. An occasional glass of wine with dinner is rarely harmful to your health and may even be beneficial, as long as you're not drinking and driving. Drink several glasses of wine or whiskey, night after night, and the story changes. As you "dose" yourself with more and more alcohol, the poisonous effects of alcohol take over, damaging your liver, heart, and digestive system and putting you at risk for cancer and other serious health problems. The more you drink, the more damage you do.

Stress and telomeres have a similar relationship. A small dose of stress does not endanger your telomeres. In fact, short-term, manageable stressors can be good for you, because they build your coping muscles. You develop skills and confidence that you can handle challenges. Physiologically, short-term stress can even boost your cells' health (a phenomenon called hormesis, or toughening). The ups and downs of daily life are usually not wearing to your telomeres. But a high dose of chronic stress that wears on for years and years will take its toll.

We now have evidence that links particular kinds of stress to shorter telomeres. These include long-term caregiving for a family member and burnout from job stress. As you may imagine, more serious traumas, both recent and in childhood, have also been linked to damaged telomeres. These traumas include rape, abuse, domestic violence, and prolonged bullying.[4]

Of course, it's not the situations themselves that produce the short telomeres; it's the stress responses that many people feel when they're in these situations. And even under these stressful circumstances, dose matters. A monthlong crisis at work can be stressful, but there's no reason to think your telomeres will take a hit. They are more robust than that; otherwise, we'd all be falling apart. (A recent review showed that there is a relationship between short-term stress and shorter telomeres, but that effect is so tiny that we don't think it will have a meaningful effect on an individual person.[5] And even if short-term stress shortens your telomeres, the effect is likely temporary, with telomeres quickly recovering their lost base pairs.) But when stress is an enduring, defining feature of your life, it can act as a slow drip of poison. The longer the stress lasts, the shorter your telomeres. It is vitally important to get out of long-term, psychologically toxic situations if it's at all possible.

But fortunately for the many of us who live with stressful situations we cannot change, that's not the whole story. **Our studies have shown that being under chronic stress does not *inevitably* lead to telomere damage.** Some of the caregivers we've studied were weathering enormous burdens without losing telomere length. These stress-resistant outliers have helped us understand that you do not necessarily have to escape difficult situations to protect your telomeres. Incredible as it sounds, you can learn to use stress as a source of positive fuel—and as a shield that can help protect your telomeres.

DON'T THREATEN YOUR TELOMERES—CHALLENGE THEM

When we looked at the data for our first caregiver study, we realized we had a mystery on our hands. Some of the caregiving mothers in the group reported less stress, and these mothers had longer telomeres. We wondered: *Why* would they feel less stress? After all, they had been caregiving for just as long as the other mothers in the group. They had a similar number of daily duties and spent just as many hours in the day performing those duties (appointments, administering injections and other treatments, managing tantrums, having to hand- or tube-feed, diaper, and bathe older disabled children).

To understand what was protecting these mothers' telomeres, we wanted to see people respond to stress in real time, before our eyes. We decided to bring more women into the lab and, essentially, stress them out. Research volunteers who arrive at our stress lab are told something like, "You're going to perform some tasks in front of two evaluators. We want you to try hard to do your best. You are going to prepare a five-minute speech and then deliver it, and perform some mental arithmetic. You can make some notes for your speech, but you will have to do all the math in your head." Sound easy? Not really, and especially not in front of an audience.

One by one, the volunteers are escorted into a testing room. Each study volunteer stands at the front of the room and faces two researchers sitting at a desk. The researchers look at the volunteer in a manner best described as stony-faced. No smiling, no nodding, no encouragement. Technically, a stony-faced expression is neutral, neither positive nor negative, but most of us are used to seeing other people smile at us, nod as we talk, or at least make an effort to seem pleasant. When compared to our usual interactions, a stony expression can come across as disapproving or strict.

The researchers explain the task, saying something like, "Please take the number 4,923 and subtract the number 17 from it, out loud. Then take your answer and subtract 17 from it, and so on, as many times as you can in the next five minutes. It is important that you perform this task quickly and accurately. We will judge you on various aspects of your performance. The clock starts now."

As each volunteer begins the math task, the researchers stare at her, pencils poised to record her answer. If she fumbles (and almost everyone fumbles), the researchers turned toward each other and whisper.

Then the volunteer goes on to her five-minute speech with the same researchers evaluating her and behaving in a similar way. If she finishes before the five minutes are up, the researchers point to the timer and say, "Please continue!" As she talks, the researchers glance at each other and slightly furrow their brows and shake their heads.

This lab stressor test, developed by Clemens Kirschbaum and Dirk Hellhammer, is a staple of psychology research, and its point is definitely *not* to test math and speech skills. Instead, it's designed to induce stress. What makes it so stressful? Mental math and on-the-fly public speaking are tricky to perform well. The most stressful element, though, is what's called social evaluative stress. Anyone who tries to perform a task in front of an audience will probably feel increased stress about their performance. When that audience appears judgmental, the stress is intensified. Even though our volunteers' physical survival was not at risk and they were safe in a clean, well-lighted university lab, this test was capable of eliciting a full-blown stress response.

We've put caregivers and noncaregivers through this protocol. We assessed their thoughts at two different times during the lab stressor: just after they'd learned what they were going to do, and just after they'd finished the two tasks. What we found was that although all the women felt *some* stress, not everyone had the *same* type of stress response. And only one kind of stress response went hand in hand with unhealthy telomeres.[6]

The Threat Response: Anxious and Ashamed—and Aging

Some of the women had what's known as a threat response to the lab stressors. The threat response is an old, evolutionary response, a kind of switch to be flipped in case of dire emergency. Basically, the threat response was designed to surge when we are face-to-face with a predator who is probably going to eat us. The response prepares our body and mind for the trauma of being attacked. As you might guess, if it keeps on happening without letup, this is *not* the response associated with telomere health.

If you already suspect that you have an exaggerated threat response to stress, don't worry. In a moment, we'll show you some lab-tested ways to convert a habitual threat response into one that is healthier for your telomeres. First, though, it's important to know what a threat response looks and feels like. Physically, the threat response causes your blood vessels to constrict so that you'll bleed less if you're wounded, but also less blood flows to your brain. Your adrenal gland releases cortisol, which gives you energizing glucose. Your vagus nerve, which runs a direct line from your brain to your viscera and normally helps you feel calm and safe, withdraws its activity. As a result, your heart rate accelerates, and your blood pressure increases. You may faint or even release your bladder. A branch of the vagus innervates the muscles of facial expression, and when that nerve isn't active, it becomes even harder for someone to interpret your facial expressions accurately. If others are wearing a similarly ambiguous expression, one that leaves lots of room for your interpretation, you in turn may view them as more hostile. You tend to freeze, you are unable to run or fight—and your hands and feet get colder, making movement more difficult.

A full-throttle threat response unleashes some uncomfortable physical reactions, but there are psychological ones, too. As you might expect, the threat response is associated with fear and anxiety. Shame, too, if you're worried about failing in front of other people. People with a strong habitual threat response tend to suffer from anticipatory worry; they imagine a bad outcome to an event that hasn't happened yet. That was exactly what happened to many of the caregivers in our lab. They felt high levels of threat—not just after they had finished the tasks but *before* the tasks had even begun. This group of caregivers became fearful and anxious when they heard the somewhat vague information about having to give a speech and do mental math. They anticipated a bad outcome, and they felt failure and shame.

As a group, our caregivers had a stronger threat response. The chronic stress of being a caregiver had made them more sensitive to a lab stressor. The ones with the strongest threat responses also had the shortest telomeres. The noncaregivers were less likely to have an exaggerated threat response, but those who did had shorter telomeres, too. Having a large anticipatory threat response—meaning that they felt threatened at the mere thought of the lab stressor before it even happened—was what mattered most.[7] Here was some vital information about how stress gets into our cells. **It's not just from experiencing a stressful event, it's also from feeling threatened by it, even if the stressful event hasn't happened yet.**

Excited and Energized: The Challenge Response

Feeling threatened is not the only way to respond to stress. It's also possible to feel a sense of challenge. People with a challenge response may feel anxious and nervous during a lab stressor test, but they also feel excited and energized. They have a "bring it on!" mentality.

Our colleague, Wendy Mendes, a health psychologist at the University of California, San Francisco (UCSF), has spent over a decade examining the body's responses to different types of stressors in the lab, and has mapped out the differences that occur in the brain, in the body, and in behavior during "good stress" compared to "bad stress." Whereas the threat response prepares you to shut down and tolerate the pain, the challenge response helps you muster your resources. Your heart rate increases, and more of your blood is oxygenated; these are positive effects that allow more blood to flow where it's needed, especially to the heart and brain. (This is the opposite of what happens when you're threatened. Then, the blood vessels constrict.) During the challenge response, your adrenal gland gives you a nice shot of cortisol to increase your energy—but then your brain quickly and firmly shuts off cortisol secretion when the stressful event is over. This is a robust, healthy kind of stress, similar to the kind you may have when you exercise. The challenge response is associated with making more accurate decisions and doing better on tasks, and is even associated with better brain aging and a reduced risk of developing dementia.[8] Athletes who have a challenge response win more often, and a study of Olympic athletes has shown that these highly successful folks have a history of seeing their life problems as challenges to be surmounted.[9]

The challenge response creates the psychological and physiological conditions for you to engage fully, perform at your best, and win. The threat response is characterized by withdrawal and defeat, as you slump in your seat or freeze, your body preparing for wounding

Figure 14: Threat versus Challenge Responses. People tend to have many thoughts and feelings when facing a stressful situation. Here are two different types of responses: One is characterized by feeling threatened, by a fear of losing, or possibly being shamed. The other is characterized by feeling challenged and confident about achieving a positive outcome.

and shame as you anticipate a bad outcome. A predominant habitual threat response can, over time, work itself into your cells and grind down your telomeres. A predominant challenge response, though, may help shield your telomeres from some of the worst effects of chronic stress.

People don't generally show responses that are *all* threat or *all* challenge. Most experience some of both. In one study, we found that it was the proportion of these responses that mattered most for telomere health. The volunteers who felt more threat than challenge had shorter telomeres. Those who saw the stressful task as more of a challenge than a threat had longer telomeres.[10]

What does this mean for you? It means you have reason to be hopeful. We do not mean to trivialize or underestimate the potential that very tough, difficult, or intractable situations have for harm

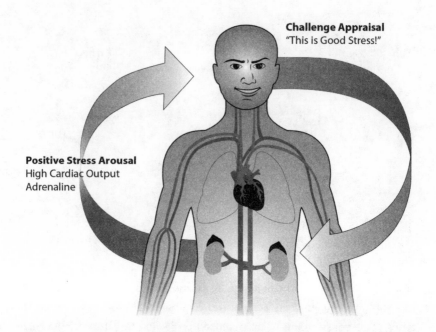

Figure 15: Positive Stress (Challenge Stress) Energizes. Our body automatically reacts to a stressful event within seconds and also reacts to our thoughts about the event. When we start to notice the stress response in our muscle tension, heart rate, and breathing, we can relabel it by saying, "This is good stress, energizing me so I can perform well!" This can help shape the body's response to be more energizing, bringing more dilation to the vessels and more blood to the brain.

to your telomeres. But when you can't control the difficult or stressful events in your life, you can still help protect your telomeres by shifting the way you view those events.

WHY DO SOME PEOPLE FEEL MORE THREAT THAN OTHERS?

Reflect on incidents in your life that have been difficult. Ask yourself: Do you tend to respond by feeling more threatened or challenged? Do you borrow trouble, feeling anticipatory threat about events that haven't happened yet—and that may not ever happen?

When you're stressed, do you feel ready for action, or do you feel like diving under the covers and hiding?

If you tend to feel more of a threat response, don't waste your time feeling bad about it. Some of us are simply wired to be more stress reactive. It has been critical to human survival for some of us to respond in a robust way to changes in our environment, and for others to be more sensitive. After all, someone's got to alert the tribe to dangers and warn the more gung-ho members against taking foolhardy risks.

Even if you weren't strongly wired at birth to feel threat, conditions in your life may have altered your natural response. Teenagers who were exposed to maltreatment when they were children respond to stressful tasks with blood-flow patterns characteristic of a threat response, experiencing vasoconstriction rather than strong blood flow out of the heart.[11] (On the other hand, people who experienced moderate adversity in childhood tend to show more of a challenge response than people who had it easy as children—more evidence that small doses of stress can be healthy, provided that resources are available to help you cope.) As we described earlier, prolonged stress can wear down emotional resources, making people more prone to feeling threatened.[12]

Either by birth or by the circumstances of your life, you may have a strong threat response. The question is: Can you learn to feel challenged instead? Research says the answer is *yes*.

DEVELOPING A CHALLENGE RESPONSE

What happens as an emotion arises? Scientists used to believe it was a more linear process—that we experience events in the world, our limbic system reacts with an emotion, like anger or fear, which causes the body to respond with an increased heart rate or sweaty palms. But it's more complicated than that. The brain is wired to *predict things ahead of time*, not just *react after things have happened.*[13] The brain uses memories of past experiences to continually anticipate what will happen next, and then corrects those predictions with both the current incoming information from the outside world, and from all the signals within our body. *Then* our brain comes up with an emotion to match all of this. Within seconds, we patch all this information together, without our awareness, and we feel some emotion.

If our "database" of past experience has a lot of shame in it, we are more likely to expect shame again. For example, if you feel high arousal and jittery, maybe from that morning's strong coffee, and if you see two people who could be talking about you, your mind may quickly cook up the emotions of shame and threat. Our emotions are not pure reactions to the world; they are our own fabricated constructions of the world.[14]

Knowing how emotions are created is powerful. Once you know this, you can have more choice over what you experience. Instead of feeling your body's stress responses and viewing them as harmful, a common experience in your brain's database, you can think about your body's arousal as a source of fuel that will help your brain work quickly and efficiently. And if you practice this enough, then eventually your brain will come to predict feelings of arousal as helpful. Even if you're one of those people whose brain is hardwired to feel more threat, you can feel that immediate instinctive survival response—and then revise the story. You can choose to feel challenged.

Sports psychologist Jim Afremow, PhD, who consults with professional and Olympic athletes, was once approached by a sprinter who was struggling with her hundred-meter time. She had already diagnosed the reason she wasn't running as well as she wanted to. "It's the stress," she said. "Before every race, my pulse races. My heart is about to jump out of my chest. You've got to help me stop it!"

Afremow laughed. "Do you really want to stop your heart?" The worst thing athletes can do, he says, is try to get rid of their stress.

"They need to think of stress as helping them get ready to perform. They need to say, 'Yes! I need this!' Instead of trying to make the butterflies in their stomach go away, athletes need to make those butterflies line up and fly in formation." In other words, they need to make the stress work for them.

The sprinter took Afremow's advice. By viewing her physical responses as tools that would help her rise to the challenge of a race, she was able to shave milliseconds off her time (a big deal for a hundred-meter runner) and set a personal record.

It sounds unbelievably simple, but research backs up this efficient method of converting threat to challenge. When research volunteers are told to interpret their body's arousal as something that will help them succeed, they have a greater challenge response. One study found that students who are encouraged to view stress in this way score higher on their GREs.[15] And when researchers put people through lab stressors, the ones who are told to think of stress as useful are able to maintain their social equilibrium. Instead of looking away, playing with their hair, or fidgeting—all signs of feeling somewhat threatened—the challenge participants make direct eye contact. Their shoulders are relaxed, and their bodies move fluidly. They feel less anxiety and shame.[16] All these benefits happened simply because people were told to think of their stress as good for them.

A challenge response doesn't make you less stressed. Your sympathetic nervous system is still highly aroused, but it is a positive arousal, putting you in a more powerful, more focused state. To channel your stress so that it gives you more good energy for an event or performance, say to yourself, "I'm excited!" or "My heart is racing and my stomach is doing cartwheels. *Fantastic*—those are the signs of a good, strong stress response." Of course, if you are under the kind of emotionally depleting stress that our caregiving mothers experienced, this language could feel too glib. Instead, talk to yourself in a gentler way. You could say, "My body's responses are trying to help me. They're designed to help me focus on the tasks at hand. They're a sign that I care." The challenge response is not a falsely chipper, gee-I'm-so-happy-that-stressful-things-are-happening-to-me attitude. It is the knowledge that even though times may be very difficult, you can shape stress to your purpose.

For those who feel addicted to "good stress"—the achievement stress involved in the constant excitement of, for example, working in a start-up company and never having downtime, know that even good stress can be overdone. It's healthy to have times when your cardiovascular system is mobilized and your psyche is primed for action. But our bodies and minds aren't built to sustain this kind of high stimulation on a consistent basis. Being able to relax, although it's been overrated as a sole source of stress management, is still necessary. We recommend that you regularly engage in an activity that brings you deep restoration. There is high-quality evidence that meditation, chanting, and other mindfulness practices can reduce stress, stimulate telomerase, and perhaps even help your telomeres to grow. See page 153 to learn more about these cell-protecting strategies.

Even in chronically stressful situations like caregiving, the stress is not a monolith or a blanket of darkness that cannot be lifted. Stress and stressful events do not live in each little moment, although they can visit. There is some freedom in each moment, because we can have a choice about how we spend this moment. We can't rewrite the past and we can't dictate what happens in the future, but we can choose where to place our attention in the moment. And although we can't always choose our immediate reactions, we can shape our subsequent responses.

Some clever studies have shown that merely anticipating a stressful event has almost the same effect on the brain and body as experiencing the stressful event.[17] When you worry about events that haven't happened yet, you're letting stress flow over its time boundaries the way a river can overflow its banks, flooding the minutes, hours, and days that could otherwise be more enjoyable. It is almost

OUR FEATHERY FRIENDS: STRESSED BIRDS, STRESSED TELOMERES

Is the stress-telomere relationship really causal? To test this, researchers have done experiments on birds. When wild European shag chicks were given water laced with the stress hormone cortisol, or were stressed out by being held, they developed shorter telomeres compared to controls.[18] Not good, since in this species, short telomeres early in life predict early death! When parrots are housed alone and can't have their usual social chats with each other, they develop shorter telomere length.[19] We know humans are sensitive to their social environments, and it seems birds are, too.

always possible to find something to worry about and therefore possible to keep the stress response engaged on an almost constant basis. When you anticipate a bad outcome before an event has even begun, you increase your dose of threat stress, and that's the last thing you need. But rather than avoiding thinking about stressful things, it's how we think about them that matters.

A SHORT PATH TO A LONG DISEASESPAN: STRESS, AGING IMMUNE CELLS, AND INFLAMMATION

It never fails. Just after you've met an important work deadline, or as you're boarding a plane for a long-overdue beach vacation, you come down with the mother of all colds: sneezing, runny nose, sore throat, fatigue. Coincidence? Probably not. While your body is actively fighting stress, your immune system can be bolstered for a time. But that effect can't last forever. Chronic stress suppresses aspects of the immune system, leaving us more vulnerable to infections, causing us to produce fewer antibodies in response to vaccinations, and making our wounds heal more slowly.[20]

There is an unsavory relationship between stress, immune suppression, and telomeres. For years, scientists were unsure just how stress, which lives in the mind, could damage the immune system. Now we have an important part of the answer: telomeres. People with chronic stress have shorter telomeres, and short telomeres can lead to prematurely aging immune cells, which means worse immune function.

Shorter Telomeres, Weaker Immune System

Certain immune cells are like SWAT teams that fight viral infections. These cells are known as T-cells, because they are stored in the thymus gland, which sits under the sternum bone in the chest. Once T-cells mature, they leave the thymus and circulate continuously throughout the body. Each T-cell has a unique receptor on its surface. The receptor acts like a searchlight on a police helicopter, sweeping the body and looking for "criminals"—cells that are either infected or cancerous. Of particular interest to aging is the type of T-cell called a CD8 cell.

But it isn't enough for the T-cell to simply spot a villainous cell. In order to complete the job, the T-cell needs to receive a second signal from a surface protein, called CD28. When the T-cell kills its target, the cell develops "memory" so that if the same virus infects the body again in the future, the T-cell can multiply into thousands and thousands of progeny cells just like itself. Together they can mount a rapid, efficient immune response against that specific virus. This is the basis of vaccination. The vaccine is typically a piece of a viral protein or a killed virus; the immunity lasts for years, since the T-cells that have responded to the initial vaccination remain in the body for a long time (sometimes for life) and are available to fight off an infection if the virus should work its way into the body again.

We have a tremendously large repertoire of T-cells, each with the capacity to recognize just one particular antigen or virus. Because we have such a huge variety of different T-cells, when we become infected with a particular virus, the few T-cells that have the correct receptor for the virus must create many progeny in order to combat the infection. During this massive process of cell division, telomerase is ratcheted up to high levels. However, it simply can't keep up with the speedy rate of telomere shortening, and eventually the telomerase response weakens to a whisper, and the telomeres in those responding T-cells keep getting shorter. So they pay for those heroic responses. When a T-cell's telomeres grow short, the cell becomes old, and it loses the CD28 surface marker that is necessary for mounting a good immune response. The body becomes like a city that's lost its budget for police helicopters and searchlights. The city looks normal from the outside, but lies vulnerable to criminal infestation. The antigens on bacteria, viruses, or cancerous cells are not cleared from the body. That's a reason people with aging cells— including the elderly and the chronically stressed—are so vulnerable to sickness, and why it's hard for them to weather diseases like the flu or pneumonia. It's partly why HIV progresses to AIDS.[21]

When telomeres in these aging T-cells are too short, even young people are more vulnerable. Sheldon Cohen, a psychologist at Carnegie Mellon University, asked young, healthy volunteers to live isolated in hotels so he could study the effects of giving them a noseful of the virus that causes the common cold. First, he measured their telomeres. The people with shorter telomeres in their immune cells, and especially in their near-senescent CD8 cells, developed colds faster, with more severe symptoms (which were measured by weighing their used tissues).[22]

What's Stress Got to Do with It?

Our CD8 T-cells (the fighters in the immune system) appear to be especially vulnerable to stress. In another of our family caregiver studies, we took blood samples from mothers who had a child with autism living at home. We found that these caregiving mothers had lower telomerase in their CD8 cells that had lost the critical CD28 surface marker, suggesting they would be in danger of developing critically short telomeres over the years. Rita Effros, an immunologist from University of California, Los Angeles, and a pioneer in understanding aging immune cells, has created "stress in a dish"—she has shown that exposing immune cells to the stress hormone cortisol dampens their levels of telomerase.[23] A compelling reason to learn how to respond to stress in a healthier way.

Shorter Telomeres, More Inflammation

Unfortunately, the news gets worse. When the telomeres of aging CD8 cells wear down, the aging cells send out proinflammatory cytokines, those protein molecules that create systemic inflammation. As the telomeres continue to shorten and the CD8 cells become fully senescent, they refuse to die and they accumulate in the blood over time. (Normally CD8 T-cells gradually die by a natural type of cell death called apoptosis. Apoptosis rids the body of old or damaged immune cells so they do not overwhelm the body or develop into the types of blood cancers called leukemias.) These senescent T-cells are the rotten apples in the barrel, with their bad effects spreading outward. They pump out slightly more inflammatory substances each year like a slow drip. If you have too many of these aging cells in your bloodstream, you're at risk for rampant infections *and* all the diseases of inflammation. Your heart, your joints, your bones, your nerves, and even your gums can become diseased. When stress makes your CD8 cells grow old, you grow old, too—no matter what your chronological age is.

Experiencing stress and pain is unavoidable. It is part and parcel of being involved with life, of loving and caring for people, caring about issues, and taking risks. Use the challenge response to protect your cells while you engage fully with life. The Renewal Lab at the end of this chapter offers some specific techniques to help you cultivate this response. The challenge response is not the only tool in your box, though. For powerful stress-relievers that are great for your telomeres, check out "Stress-Reducing Techniques Shown to Boost Telomere Maintenance" at the end of Part Two. And if stress tends to lead you into destructive thinking patterns—maybe you suppress painful thoughts or ruminate excessively about them, or perhaps you begin to anticipate negative responses from other people—turn to the next chapter. We'll help you protect your telomeres from this harmful thinking.

TELOMERE TIPS

- Your telomeres don't sweat the small stuff. Toxic stress, on the other hand, is something to watch for. Toxic stress is severe stress that lasts for years. Toxic stress can dampen down telomerase and shorten telomeres.
- Short telomeres create sluggish immune function and make you vulnerable even to catching the common cold.
- Short telomeres promote inflammation (particularly in the CD8 T-cells), and the slow rise of inflammation leads to degeneration of our tissues and diseases of aging.
- We cannot rid ourselves of stress, but approaching stressful events with a challenge mentality can help promote protective stress resilience in body and mind.

RENEWAL LAB

REDUCE "EGO THREAT" STRESS

If you feel that an important aspect of your identity is on the line, you are probably going to feel a strong threat response. This is why a final exam can be so stressful if your main identity is as a "good student," or why a sports competition can feel terrifying if you strongly identify as an athlete. If you do poorly, you don't just suffer a bad grade or a loss. The experience takes a bite out of your sense of self-worth. A challenge to your identity leads to threat stress, which can lead to poor performance, which can wound your identity. It's a vicious cycle, one that may have a negative effect on your telomeres. Break the cycle by reminding yourself that your identity runs wide and deep:

Instructions for defusing ego threat: Think of a stressful situation. Now in your mind or on a piece of paper, make a list of the things that you value (it's best to choose things unrelated to the stressful situation). For example, you may think about some social roles that are important to you (being a parent, good worker, community member, etc.) or values you believe are particularly important (such as your religious beliefs, community service). Next, think about a specific time in your life when one of these roles or values was particularly salient for you.

There are many studies documenting this effect; typically in these studies, volunteers are asked to write for ten minutes about their personal values. This small manipulation (called value affirmation) reduces stress responses in the lab, and in real life, and helps people engage in stressful tasks with a challenge mind-set.[24] Identifying values translates into better performance and higher grades on science tests.[25] It activates the reward area of the brain that may help buffer stress reactions.[26]

The next time a threat looms, pause and list what's most important to you. One caregiving mother we know stops and reminds herself that one of her highest priorities is helping her son who has autism, which seems to absorb her tension and protect her from caring about what other people think. When he has a meltdown in a public space, she ignores the judgmental stares from other people and simply does what her son needs. "It's like I'm in a protective bubble," she says. "It's a lot less stressful in there." When you see just how broad your values run, you validate your sense of self-worth, so there's less of your identity riding on the outcome of a single event.

DISTANCING

Create some space between your feeling self and your thinking self. Researchers Ozlem Ayduk and Ethan Kross and their colleagues have conducted several lab studies to manipulate the emotional stress response, in order to see what amps it up and what allows emotions to dissipate quickly. They've discovered that by distancing your thoughts from your emotions, you can convert a threat response into a positive feeling of challenge. Below are the methods Ayduk and Kross have identified to create this distance:

Linguistic self-distancing. Think about an upcoming stressful task using the third person, as in "What is making Liz nervous?" Thinking in the third person "puts you in the audience," so to speak, or makes you a fly on the wall. You don't feel so caught up in the drama. Moreover, research shows that frequent self referencing ("I," "me," and "mine") is a sign of being self-focused and is related to feeling more negative emotions. Ayduk and Kross have found that thinking in the third person and not using "I" leads people to feel less threatened, anxious, and ashamed, and to engage in less rumination. They perform better at stressful tasks, and raters view them as more confident.[27]

Time distancing. Think about the immediate future, and you will have a bigger emotional response than if you take a longer-term view. Next time you are in the grip of a stressful event, ask yourself, *In ten years, will this event still have an impact on me?* In studies, people asked to pose this question to themselves had more challenge thoughts. Recognizing the impermanence of an event helps you get over it faster.

Visual self-distancing. Distancing is a trick you can play on the threat response after the fact. If you have experienced a stressful event that you still feel emotional about, visual distancing allows you to emotionally process it in a way that will help put it to rest. Rather than just relive the event straight up, which can induce the same emotions you felt in those moments, *step back and view the event from afar, as if it's happening in a movie that you're watching.* That way, you won't reexperience the event in your emotional brain. Instead, you'll view it with greater separation and clarity. Distancing takes some of the power away from a negative memory. This technique is also known as cognitive defusion, and it's been shown to immediately reduce the brain's neural stress response,[28] probably because it activates the brain's more reflective, analytical areas instead of its emotional ones. Here is a modified version of the script Ayduk and Kross use to help their research volunteers create distancing (we combined visual, linguistic, and time distancing):[29]

> *Instructions for distancing:* Close your eyes. Go back to the time and place of the emotional experience and see the scene in

your mind's eye. Now take a few steps back. Move away from the situation to a point where you can now watch the event unfold from a distance and see yourself in the event, the distant you. Now watch the experience unfold as if it were happening to the distant you all over again. Observe your distant self. As you continue to watch the situation unfold for your distant self, try to understand his [or her] feelings. Why did he [or she] have those feelings? What were the causes and reasons? Ask yourself, "Will this situation affect me in ten years?"

If you suffer from retrospective stress—if you feel a lot of negative emotions and shame after an event is over—the visual distancing strategy can be especially useful. You can also try this strategy while you're actually in the stressful moment. By mentally stepping outside your body, you can bypass its sense of imminent threat and attack.

CHAPTER FIVE

Mind Your Telomeres: Negative Thinking, Resilient Thinking

We are largely unaware of the mental chatter in our minds and how it affects us. Certain thought patterns appear to be unhealthy for telomeres. These include thought suppression and rumination as well as the negative thinking that characterizes hostility and pessimism. We can't totally change our automatic responses—some of us are born ruminators or pessimists—but we can learn how to keep these automatic patterns from hurting us and even find humor in them. Here we invite you to become more aware of your habits of mind. Learning about your style of thinking can be surprising and empowering. To see what your own tendencies are, take the personality assessment at the end of this chapter (page 128).

One day several years ago, Redford Williams came home from a difficult day at the office and headed for the kitchen. Then he stopped. There was a pile of catalogs sitting on the counter—a pile that his wife, Virginia, had agreed to get rid of the day before. Yet there Virginia stood, calmly stirring a pot on the stove. The catalogs remained exactly where she'd left them.

Redford exploded. "Get the damn catalogs off the counter!" he ordered. It was the first thing he'd said to her since walking in the door.

What was he thinking? It's a natural question when we hear about bewildering, out-of-proportion hostility like this. Because Redford Williams is now a renowned professor of psychology and neuro-science at Duke University and an expert in anger management, he can provide some answers. "I was thinking that I was exhausted, surprised, and angry. I was thinking that Virginia was being lazy, and that she was deliberately avoiding a task she'd promised to do," he said. "I was impugning her motives." He discovered later that Virginia had not moved the catalogs because she was busy cooking him a meal that would be good for his heart.

Scientists are learning that certain thought patterns are unhealthy for telomeres. Cynical hostility, which is characterized in part by the kind of suspicious, angry thoughts that gripped Williams when he saw a less than perfectly tidy kitchen, is linked to shorter telomeres. So is pessimism. Other thought patterns, including mind wandering, rumination, and thought suppression, may also lead to telomere damage.

These thought patterns, unfortunately, can be automatic and hard to change. Some of us are born cynics or pessimists; some of us have been ruminating about our problems practically since we were old enough to talk. In this chapter, we'll describe each of these automatic patterns, but you'll also discover that you can learn to laugh at your negative thoughts and keep them from hurting you so much.

CYNICAL HOSTILITY

In the 1970s, the best-selling book *Type A Behavior and Your Heart* made *type A personality* a household term. The book claimed that type A behavior—characterized by hard-charging impatience, an emphasis on personal achievement, and hostility toward others— was a risk factor for heart disease.[1] You still see the idea of type A lingering in online assessments and casual conversation. ("Oh, I hate to stand in long lines—I'm so type A.") Actually, subsequent research showed that being a quick-on-the-draw high achiever is not necessarily harmful to your health. It's the hostility component of type A that is so damaging.

Cynical hostility is defined by an emotional style of high anger and frequent thoughts that other people cannot be trusted. A person with hostility doesn't just think, "I hate to stand in long lines at the grocery store." A person with hostility thinks, "That other shopper deliberately sped up and beat me to my rightful position in the line!"—and either seethes or makes a nasty expression or comment to the unsuspecting person who is standing in front of him. People who score high on measures of cynical hostility often cope passively by eating, drinking, and smoking more. They tend to get more cardiovascular disease, metabolic disease,[2] and often die at younger ages.[3]

They also have shorter telomeres. In a study of British civil servants, men who scored high on measures of cynical hostility had shorter telomeres than men whose hostility scores were low. The most hostile men were 30 percent more likely to have a combination of short telomeres and high telomerase—a profile that is worrisome, because it seems to reflect unsuccessful attempts of telomerase to protect telomeres when they are too short.[4]

The men who had this vulnerable cell aging profile had the opposite of a healthy response to stress. Ideally, your body responds to stress with a spike in cortisol and blood pressure, followed by a quick return to normal levels. You're prepared to meet whatever challenge is facing you, and then you recover. When these men were exposed to stress, their diastolic blood pressure and cortisol levels were blunted, a sign that their stress response was, basically, broken from overuse. Their systolic blood pressure increased, but instead of returning to normal levels after the stressful event was over, it stayed elevated for a long time afterward. The men also had few of the resources that usually buffer people from stress. In addition to the greater hostility, they had fewer social connections and less optimism, for example.[5] In terms of their physical and psychosocial health, these men were highly vulnerable to an early diseasespan.

Women tend to have lower hostility, and it is less related to heart disease for them, but there are other psychological culprits affecting women's health, such as depression.[6]

PESSIMISM

One of the brain's chief jobs is to predict the future. The brain is constantly scanning the environment and comparing it to past experience, looking for upcoming threats to your safety. Some people have brains that are faster to spot danger. Even in ambiguous or neutral situations, these people tend to think, "Something bad is going to happen here." These folks are the first to prepare for a worst-case scenario, the first to expect a bad outcome. In other words, they're pessimists.

I (Elissa) am reminded of pessimism when I am out hiking with my friend Jamie. I see off-trail paths as an adventure; she sees them as potential for poison oak. When we see a house in the woods or in the middle of nowhere, I feel some delight and anticipation. Someone may invite us in for tea! Or maybe we'll at least get a smile and a hello if someone steps out onto the porch. But Jamie has a different set of thoughts. She's sure that if someone steps out onto the porch, it will be with a furrowed brow, gruff words, and maybe even a rifle. Jamie has a more pessimistic style of thought.

When our research team conducted a study on pessimism and telomere length, we found that people who scored high on a pessimism inventory had shorter telomeres.[7] This was a small study, of about thirty-five women, but similar results have been found in other studies, including a study of over one thousand men.[8] It also fits with a large body of evidence that pessimism is a risk factor for poor health. When pessimists develop one of the diseases of aging, like cancer or heart disease, the disease tends to progress faster. And, like cynically hostile people—and people with short telomeres generally—they tend to die earlier.

We already know that people who feel threatened by stress tend to have shorter telomeres than people who feel challenged by it. Pessimists, by definition, feel more threatened by stressful situations. They are more likely to think they won't do well, that they can't handle the problem, and that the problem is going to linger. They tend *not* to get pumped up about a challenge.

Although some people are born pessimists, some kinds of pessimism are forged by early environments in which a child learns to expect deprivation, violence, or distress. In these situations, pessimism can be seen as a healthy adaption, a protection against the pain of repeated disappointment.

MIND WANDERING

As you sit, holding this book or your e-reader in your hands, are you thinking about what you're reading? If you are thinking about something else, are your thoughts pleasant, unpleasant, or neutral? And how happy are you, right now?

Harvard psychologists Matthew Killingsworth and Daniel Gilbert used a "track your happiness" iPhone app to ask thousands of people questions like these. At random times across the day, the app prompts people to respond to similar questions about what activity they are engaged in, what their minds are doing, and how happy they are.

As the data came in, Killingsworth and Gilbert discovered that we spend half of the day thinking about something other than what we're doing. This is true almost no matter what activity is at hand. Having sex, engaging in conversation, or exercising are the activities that produce the least mind wandering, but even these have a 30 percent mind-wandering rate. "The human mind is a wandering mind," they concluded. Emphasis on "human": they noted that we are alone among the animals in our ability to think about something that's not happening right now.[9] This power of language lets us plan, reflect, and dream—but it's a power that comes with a price tag.

The iPhone mind-wandering study showed that when people are

not thinking about what they're doing, they're just not as happy as when they're engaged. As Gilbert and Killingsworth also observed, "A wandering mind is an unhappy mind." In particular, *negative* mind wandering (thinking negative thoughts, or wishing you were somewhere else) was more likely to lead to unhappiness in their next moments—no surprise there. (To gauge how often your own mind wanders, you can download the app at https://www.trackyourhappiness.org.)

Together with our colleague Eli Puterman, we studied close to 250 healthy, low-stress women who ranged from fifty-five to sixty-five years old, and we assessed their tendency to mind-wander. We asked them two questions to assess their presence in the moment and negative mind wandering:

How often in the past week have you had moments when you felt totally focused or engaged in doing what you were doing at the moment?

How often in the past week have you had any moments when you felt you didn't want to be where you were, or doing what you were doing at the moment?

Then we measured the women's telomeres. The women with the highest levels of self-reported mind wandering (which we defined as low present-oriented focus along with wanting to be somewhere else) had telomeres that were shorter by around two hundred base pairs.[10] This was regardless of how much stress they had in their lives. That's why it's a good habit to notice if you are having thoughts of wanting to be somewhere else. That thought reveals an internal conflict that creates unhappiness. This type of negative mind wandering is the antithesis of a mindful state. As Jon Kabat-Zinn, founder of the worldwide program Mindfulness Based Stress Reduction (MBSR), has said, "When we let go of wanting something else to happen in this moment, we are taking a profound step toward being able to encounter what is here now."[11]

Splitting your attention by multitasking is a low-grade source of noxious stress, even if you are not aware of it. We naturally mind-wander much of the time, and some kinds of mind wandering can be creative. But when you are thinking negative thoughts about the past, you are more likely to be unhappy, and you may possibly even experience higher levels of resting stress hormones.[12] It's becoming increasingly clear that *negative* mind wandering may be an invisible source of strife.

UNITASKING

We all have pressure on our limited attention these days and are inclined to multitask, to check e-mail, to use our time efficiently. It turns out the most efficient use of time is to do one thing and to pay full attention to it. This "unitasking," sometimes termed "flow," is also the most satisfying way to spend moments. We allow ourselves to be content and absorbed. When I (Elissa) have a day of meetings, I can easily and frenetically split myself between giving divided bits of attention to the meeting, my phone, e-mail, and intrusive thoughts about what else I need to be doing, or I can decide to be focused fully on the person in front of me. The latter is a simple pleasure, and the person in front of me has a different experience as well.

And I (Liz) felt the same contrasting sets of pulls and tugs on my attention when I was an active research scientist and mother while also serving an administrative position as the chair of my department at UCSF. On a day when I would allow myself to become absorbed in my lab doing experimental manipulations with molecules and cells in tiny test tubes, hours of productive work would fly by before I was aware of their passing. Any weekend at home just spending time with the family seemed to finish almost as soon as it had begun. Those times felt very different from juggling many kinds of time-constrained work duties. Of course, sometimes such tight schedules with multitasking cannot be avoided. But whatever you are doing, whether it is in the form of "flow" or various rapidly transitioning activities, you can try to eliminate other distractions, and be fully present, at least for part of the day.

RUMINATION

Rumination is the act of rehashing your problems over and over. It's seductive. Rumination's siren call sounds something like this: *If you keep mulling things over, if you think some more about an unresolved issue or why a bad thing happened to you, you'll have some kind of cognitive breakthrough. You'll solve problems, you'll find relief!* But rumination only *looks* like the act of problem solving. Being caught in rumination is more like getting sucked into a whirlpool that hurtles you through increasingly negative, self-critical thoughts. When you ruminate, you are actually less effective at solving problems, and you feel much, much worse.

How do you tell rumination from harmless reflection? Reflection is the natural curious, introspective, or philosophical analysis about why things happen a certain way. Reflection may cause you some healthy discomfort, especially if you are thinking about something you wish you hadn't done. But rumination feels *awful*. You can't stop yourself, even if you try. And it doesn't lead to a solution, only to more ruminating.

If for some reason you wanted to prolong the bad effects of stress long after a difficult event was past, rumination would be an effective way to do it. When you ruminate, stress sticks around in the body long after the reason for the stress is over, in the form of prolonged high blood pressure, elevated heart rate, and higher levels of cortisol. Your vagus nerve, which helps you feel calm and keeps your heart and digestive system steady, withdraws its activity—and it remains withdrawn long after the stressor is over. In one of our most recent studies, we examined daily stress responses in healthy women who were family caregivers. The more the women ruminated after a stressful event, the lower the telomerase in their aging CD8 cells, the crucial immune cells that send out proinflammatory signals when they are damaged. People who ruminate experience more depression and anxiety,[13] which are in turn associated with shorter telomeres.

THOUGHT SUPPRESSION

The final dangerous thought pattern we'll describe is actually a kind of antithought. It's a process called thought suppression, the attempt to push away unwanted thoughts and feelings.

The late Daniel Wegener, a Harvard social psychologist, was reading one day when he came across this line from the great nineteenth-century Russian writer Fyodor Dostoevsky: "Try to pose for yourself this task: not to think of a polar bear, and you will see that the cursed thing will come to mind every minute."[14]

Wegener, feeling that this idea rang true, decided to put it to the test. Through a series of experiments, Wegener identified a phenomenon he called ironic error, meaning that the more forcefully you push thoughts away, the louder they will call out for your attention. That's because suppressing a thought is hard work for your mind. It has to constantly monitor your mental activity for the forbidden item: *Is there a polar bear around here anywhere?* The brain can't sustain that work of monitoring. It fatigues. You try to push the polar bear behind an ice floe, and it pops back up, poking its head above the water and bringing a few friends along for good measure. You get *more* thoughts of polar bears than if you didn't try to suppress them in the first place. Ironic error is one reason smokers who are trying to quit will constantly think of cigarettes and why dieters, trying desperately not to think of food, are tortured by images of sweet Frappuccinos.

Ironic error may also be harmful to telomeres. We know that chronic stress can shorten telomeres—but if we try to manage our stressful thoughts by sinking the bad thoughts into the deepest waters of our subconscious, it may backfire. The chronically stressed brain's resources are already taxed (we call this cognitive load), making it even harder to successfully suppress thoughts. Instead of less stress, we get *more*. A classic example of suppression's dark power comes from people with post-traumatic stress disorder (PTSD), who—understandably—don't want to remember events that have caused them terrible distress. But their ghastly memories barge into their daily lives in an unexpected, jarring way or enter their dreams at night. Often they will judge themselves harshly for letting the intrusive thought into their minds—for not being strong enough to hold it back—and for having an emotional response to the thought.

Take a moment to absorb the links here. We push away our bad feelings, which inevitably roar back, and then we feel bad, and *then* we feel bad about feeling bad. That additional layer of negative judgment—the layer of feeling bad about feeling bad—can be like a heavy blanket that smothers that last bit of energy you had available for coping. It's one reason people fall into a serious depressive state. In one small study, greater avoidance of negative feelings and thoughts was associated with shorter telomeres.[15] Avoidance alone is probably not enough to shorten telomeres. But as you'll see in the next chapter, there's a body of evidence showing that untreated clinical depression is extremely bad for telomeres. In short: Thought suppression is a royal road to chronic stress arousal and depression, both of which shorten your telomeres.

THE ANATOMY OF A STRESSFUL DAY

In a recent study, we followed mothers caring for a child with an autism spectrum disorder. We wanted to understand the emotional anatomy of their days. Not surprisingly, the caregivers woke with more dread about the day than a control group of mothers with typical children. As the day unfolded, they viewed its stressful events as more threatening. The caregiving mothers ruminated more about the stressful things that had happened. They also reported more negative mind wandering. It appears that the chronic stress of caregiving creates a **hyperreactive stress syndrome**, in which stressful events are more often anticipated, worried about, overreacted to, or ruminated upon.

When we looked at these caregivers' cells, we found that telomerase was significantly lower in their aged CD8 cells. And for all the women in the study, negative thinking was associated with the lower telomerase. On the positive side, there were many caregivers who awoke with joy, who had a challenge response to stress, and who managed to avoid rumination—and these habits were all associated with higher telomerase.

RESILIENT THINKING

If you suffer from any of the painful mental habits we've just described (pessimism, rumination, negative mind wandering, and the thinking that characterizes cynical hostility) you probably want to make some changes. But it's unlikely that you'll end negative thinking just by ordering yourself to stop. People who scold themselves about needing to change their thoughts remind us of the *Seinfeld* episode when Frank Costanza, worked into a lather about the seating arrangements in George's car, raises his hands in the air and shouts, "Serenity now! Serenity now!" Frank explains that this is what he's supposed to say to calm himself when his blood pressure gets too high. George gazes pointedly into the rearview mirror at his father, who is red-faced and practically foaming at the mouth, the very opposite of serene.

"Are you supposed to *yell* it?" he asks.

Yelling at yourself doesn't work. For one thing, personality traits such as cynical hostility and pessimism have a genetic component—they're baked in. And if you had a lot of trauma in your childhood, you may have negative thoughts frequently. Those thoughts are life-long habits, and it's possible they won't ever completely go away. So scolding yourself is unlikely to be effective. Fortunately, you can shield yourself from some of the effects of negative thought patterns by using resilient thinking.

Resilient thinking is encompassed in a new generation of therapies based on acceptance and mindfulness. These therapies don't try to alter your thoughts. Instead, they help you change your relationship to them. You don't need to believe your negative thoughts, or act on them, or have a lot of bad feelings because the thoughts crossed your mind. Below are some suggestions for responding to negative thought patterns in a more resilient way. These suggestions will help you feel better—and we believe, based on preliminary clinical trials that have been conducted so far, that improving your stress resilience is good for your cellular health in general.

Thought Awareness: Loosen the Grip of Negative Thought Patterns

The negative thought patterns we've described here are automatic, exaggerated—and controlling. They take over your mind; it's as if they tie a blindfold around your brain so that you can't see what is really going on around you. When your negative thought patterns are in control, you really believe that your wife is lazy; you can't see that she is working hard to make sure you have a healthy dinner. You believe that the stranger will come out of the house with a rifle; you can't see how exaggerated this scenario is. But when you become more aware of your thoughts, you take off the blindfold. You don't necessarily stop the thoughts, but you have more clarity.

Activities that directly promote better thought awareness include most types of meditation, especially mindfulness meditation, along with most forms of mind-body exercises. Even long-distance running, with its repetitive footfalls, can help with thought awareness and present orientation. You can notice the rhythm you create when your foot hits the ground, notice details of the trees and leaves you pass, notice your thoughts passing. Engaging in any kind of mind-body practice regularly allows you to be less focused on negative thoughts about yourself; you get better at noticing your surroundings and other people. And in times of reactivity, you are able to notice you are experiencing negative thoughts, and they dissolve sooner. Thought awareness promotes stress resilience.

To become aware of your thoughts, close your eyes, take some relaxed breaths, and focus on the movie screen of your mind. Take a mental step back and watch your thoughts go by, as if you're watching traffic on a busy street. For some of us, that street is like the New Jersey Turnpike during a thunderstorm—slick, crowded, and heart-thumpingly fast. That's fine. As you become more aware of your thoughts, including the ones that are distressing, you can label them, accept them, and even laugh at them. ("Oh, I'm criticizing myself again. I do that so often it's funny.") Instead of pushing your thoughts under the surface or letting them control your behavior, you let the negative thoughts pass by.

Thought awareness can reduce rumination.[16] It can help with automatic negative thinking by putting some distance between your instinctive thought and your reaction to that thought. You realize that you don't have to follow the story line inside your head—because, as you'll notice, the story line doesn't usually lead to productive thinking. We apparently have around sixty-five thousand thoughts a day. We're not really in control of *generating* our thoughts; they come no matter what we do. And this includes thoughts we would never invite in. But when you practice thought awareness, you notice that about 90 percent of your thoughts are repeats of thoughts that came before. You feel less compelled to grab on to them and take them wherever they lead you. They're just not worth following. With time, you learn to encounter your own ruminations or problematic thoughts and say, "That's just a thought. It'll fade." And that is a secret about the human mind: We don't need to believe everything our thoughts tell us. (As the wise bumper sticker says, "Don't believe everything you think.") The only thing we can be sure of is that our thoughts are constantly changing. Thought awareness helps us perceive the truth of this statement.

I (Liz) went on a mindfulness meditation retreat several years ago in order to learn about and experience this technique, since some of my collaborative studies on telomeres involved meditation interventions. With other interested scientists, and psychologists, I spent a week at a quiet location in southern California to learn from Alan Wallace, an experienced teacher of Tibetan meditation techniques. As a newcomer to mindfulness, I was surprised to learn how much emphasis was placed on training the mind to focus its attention. I found that the mindfulness meditation techniques produced a calm mind, along with pleasant, unbidden feelings like gratitude.

Now, years later, the ability to focus better on whatever is at hand has stayed with me. To keep this ability topped up, I use micro-meditations at times that could otherwise leave me bored, antsy, or impatient: when I'm waiting for the airplane to take off; during a shuttle ride in San Francisco en route to a meeting; waiting for my computer to boot up; or even just waiting for the microwave to heat a cup of tea.

Next time you notice that unwelcome thoughts are racing through your head, you might want to try this: *Close your eyes. Let yourself breathe normally, but pay attention to your breath. When thoughts come into your mind, imagine you are simply a witness to them and watch them gently waft away. Try not to judge the thoughts or yourself for having the thoughts. Bring your attention back to your breath, focusing on the natural feel of it as you breathe in and breathe out.*

With practice, the thoughts that are buzzing in your mind will settle down, and you'll be in a more focused state. Picture your mind like a snow globe. Minds are often in an unsettled state, and the globe is cloudy with thoughts. But taking a mini-meditation break allows the thoughts to eventually settle, allowing you more mental clarity. You won't be at the mercy of following your thoughts.

Of course, it is wonderful if you can practice for a longer time, or to attend a mindfulness retreat to learn this new skill more easily.

But don't let the perfect be the enemy of the good. Shorter periods of mindfulness will also help you develop thought awareness and reduce the power of your negative thought patterns.

Mindfulness Training, Purpose in Life, and Healthier Telomeres

In one of the most dramatic and comprehensive studies of meditation ever performed, experienced meditators headed up to the Colorado Rockies for a mountain retreat with Buddhist teacher Alan Wallace. For three months, they followed an intense practice of concentration meditation aimed to cultivate a relaxed, vivid, and stable attentional focus. The meditators also engaged in practices designed to foster beneficial aspirations for themselves and others, such as compassion.[17] The meditators also put up with a lot of experimentation, including blood draws. The intrepid University of California, Davis, researcher Clifford Saron and his colleagues decided to measure the meditators' telomerase, too, so they built a wet lab right on the mountain, complete with a refrigerated centrifuge and a dry ice freezer for storing the meditators' cells at the necessary temperature of eighty degrees below zero degrees Celsius—meaning that they had to haul five thousand pounds of dry ice up the mountain over the course of the project.

The results were what you might expect from three months sitting in a beautiful locale, listening to an inspiring teacher, and meditating among like-minded indviduals every day. After the retreat, the meditators felt better—less anxious, more resilient, empathetic. They had longer sustained focus and could better inhibit their habitual responses.[18] The researchers checked in with the meditators five months after the retreat, and these effects were still strong. They

114

found that the enhanced ability to inhibit responses, gained from the retreat, predicted the longer-term improvements in emotional wellbeing.[19] A control group of experienced meditators who waited back at home for their turn on the mountain (but who were flown to the retreat center for testing) did not experience these effects until they went on the retreat themselves.

The meditators also experienced an increased sense of purpose in life. When you have a sense of purpose, you wake up in the morning with a sense of mission, and it's easier to make decisions and plans. In a study led by neuroscientist Richard Davidson of the University of Wisconsin, volunteers were exposed to distressing pictures, which typically enhances one's startle response to loud noise. The eyeblink startle response reflects an automatic defense response in the brain. The people with the strongest sense of purpose in life had a more resilient stress response, less reactivity, and faster recovery of their eyeblink startle response.[20]

Stronger feelings of life purpose are also related to reduced risk of stroke and improved functioning of immune cells.[21] Life purpose is even linked to less belly fat and lower insulin sensitivity.[22] In addition, having a higher purpose in life may inspire us to take better care of ourselves. People with greater purpose tend to get more lifesaving tests to detect early disease (such as prostate exams and mammograms), and when they do get sick, they stay in the hospital fewer days.[23] The writer Leo Rosten said, "The purpose of life is not to be happy—but to matter, to be productive, to be useful, to have it make some difference that you lived at all." But it doesn't have to be a competition between being happy and being productive with purpose—they come together.

Life purpose is what brings us eudaemonic happiness, the healthy feeling that we are involved in something bigger than ourselves. Eudaemonic happiness is not the transitory happiness we experience when eating or buying something we really want; it is enduring wellbeing. A strong sense of our values and purpose can serve as a bedrock foundation that helps us feel stability through life events, those earthquakes both minor and major. In hard times, we can bring them to mind over and over again. They may even protect us from threat stress at an unconscious automatic level. With a strong sense of purpose, the vicissitudes of life, including both joy and sorrow, can then fit more easily into a meaningful context or container.

What about cell aging? Saron had used the blood draws and wet lab to spin, separate, and save the white blood cells for later lab analysis by Liz and our colleague Jue Lin, who examined the meditators' telomerase activity. (Back then, we didn't think telomeres could change quickly, so we did not measure them in studies that followed people for only a few months.) Tonya Jacobs carefully analyzed telomerase in relation to self-reported psychological changes in wellbeing, such as purpose in life. Overall, the retreat group had 30 percent more telomerase than the group that had been wait-listed. And the more that the meditators improved on scores of purpose in life, the higher their telomerase.[24] Meditation, if it is of interest to you, is obviously one important way to enhance your own purpose in life. There are innumerable ways to achieve a greater sense of purpose, and which one you choose depends on what is most meaningful to you.

A New Purpose During Retirement? The Experience Corps

Imagine you've been retired for years. You have your routine, and you know what to expect each day. Then you are approached and asked to tutor an at-risk child in your neighborhood. What would you say? What is it like for someone not used to a daily job anymore, and certainly not used to working in a low-income school with little kids? So just what does happen when retired people join a tutoring program, volunteering fifteen hours a week?

The Experience Corps is a remarkable program that matches retired men and women as tutors in public schools for low-income, urban, young children. It is also a high-intensity volunteer experience and comes with its own stresses. A group of gerontology researchers wanted to see if this intergenerational program could improve health for all involved, so they have been examining the benefits of the program for both the children and the adults. So far the results are profound.

First, let's take a closer look at the volunteers' experiences of stress. Many of them were interviewed about the stresses and rewards of volunteering. They dealt with the kids' behavioral issues and sometimes couldn't get to their lesson. They saw the children's personal problems, sometimes including parental neglect, up close. They didn't always get along with the teachers. However, the rewards were numerous, and on balance the benefits outweighed the stressful aspects. They enjoyed helping the children and seeing them improve and developed special relationships.[25] This sounds like a form of positive stress!

To examine effects on health, researchers built a controlled trial into the Experience Corps, randomizing older adults to either volunteering in the Corps or a control condition. Two years later, the volunteers felt more "generative" (more accomplished from helping others).[26] The volunteers had some physiological transformations as well: While the control group had declines in brain volume (cortex and hippocampus), the volunteers had increases, especially the men. The men showed a reversal of three years of aging over two years of volunteering. This increase means better brain function—the larger the brain-volume increase, the better their increase in memory.[27] These increases in wellbeing and brain volume remind us that "life shrinks or expands in proportion to one's courage," as the writer Anaïs Nin has said.

A TELOMERE-HEALTHY PERSONALITY TRAIT

Personality traits like cynical hostility and pessimism may damage your telomeres, but there's one personality trait that appears to be good for them: conscientiousness. Conscientious people are organized, persistent, and task oriented; they work hard toward long-term goals—and their telomeres tend to be longer.[28] In one study, teachers were asked to rank their young students according to their conscientiousness. Forty years later, the students who'd scored highest on conscientiousness had longer telomeres than the ones who were the least conscientious.[29] This finding is important, because conscientiousness is the personality trait that is the most consistent predictor of longevity.[30]

Part of conscientiousness is having good impulse control, being able to delay the lure of immediately rewarding (and often dangerous) things like overspending money, driving too fast, excess eating, or alcohol use. Having high levels of impulsivity is associated with shorter telomeres as well.[31]

Conscientiousness in childhood predicts longevity decades later, and in a study of Medicare patients, those with high self-discipline lived 34 percent longer than their less conscientious counterparts.[32] Perhaps that's because conscientious people are better able to control impulses, engage in healthy daily behaviors, and follow medical advice. They also tend to have healthier relationships and find better work environments, all of which mutually reinforce wellbeing and thriving.[33]

Trade Pain for Self-Compassion

Another technique for resilient thinking is self-compassion. Self-compassion is nothing more than kindness toward yourself, the knowledge that you are not alone in your suffering, and the ability to turn toward and face difficult emotions without getting lost in them. Instead of beating yourself up, you treat yourself with the same warmth and understanding you'd extend to a friend.

To gauge your self-compassion, answer these questions, based on Kristin Neff's Self-Compassion Scale:[34] Do you try to be patient and tolerant toward aspects of your personality you don't like? When something painful happens, do you try to take a balanced view of the issue? Do you remind yourself that everybody has flaws and that you are not alone? Do you give yourself the care you need? Yes answers indicate that you're high in self-compassion, and you probably recover quickly from most stresses.

Now try these questions: When you fail at something important to you, do you berate yourself? Do you become consumed by feelings of inadequacy? Are you judgmental about your flaws? Do you feel isolated and alone, separate from other people?

If you've answered yes to these, it's a sign that you struggle to feel compassionate toward yourself. Self-compassion is a skill you can develop. And it's a skill that will help you develop a resilient response to your negative thoughts. (See the Renewal Lab on page 122 for a few ideas.)

When people who are high in self-compassion have a flood of negative thoughts and feelings, they do things differently from the rest of us. They don't criticize themselves for having faults. They can observe their negative thoughts without getting swept up in them. This means that they don't have to push away negative feelings; they just let those feelings happen and then fade. This kind attitude has positive effects on their health. People high in self-compassion react to stress with lower levels of stress hormones,[35] and they have less anxiety and depression.[36]

You may be objecting to the idea of self-compassion. Some people think it is more honest and more honorable to be self-critical. Of course, it's wise to have an accurate sense of your strengths and shortcomings, but that's different from judging yourself harshly. It's different from carping at yourself when you think you're not measuring up to the competition. Self-criticism cuts like a knife. It hurts, and those invisible knife wounds don't make you stronger or better. In fact, self-criticism is a particularly painful form of self-pity, not self-improvement.

Self-compassion *is* self-improving, because it cultivates the inner strength to cope with the troubles of life. By teaching us to rely on ourselves for encouragement and support, self-compassion makes us more resilient. Depending on other people to make us feel good about ourselves is fraught with peril. When we need other people to think well of us, the thought of their disapproval is so painful that we try to beat them to it—and that's when we jump to criticize ourselves. We cannot overrely on others for comfort. Developing self-compassion is not weak or wimpy at all. It is self-reliance, and a part of stress resilience.

Wake Up Joyfully

We have found that women who wake up with feelings of joy have more telomerase in their CD8 immune cells, and their waking cortisol peak is less exaggerated than women who wake up without joy or with dread. We don't know if this is causal, of course, but let's hedge our bets and talk about those first moments of waking. They can shape the rest of your day. Regardless of which day of your life it happens to be, you can start the day with gratitude. Upon waking—and before mentally jumping into your to-do list—see what it feels like to think "I am alive!" and welcome in the day. Even though you can't know or control what the future holds, you can turn your attention to the beauty of having a fresh new day and acknowledge some small thing you are grateful for.

I (Elissa) was struck by hearing how the fourteenth Dalai Lama wakes up: "Every day, think as you wake up, today I am fortunate to be alive, I have a precious human life, I am not going to waste it." It's too easy to never have this thought and to miss this life-affirming perspective.

As you've now read, there are many ways to promote stress resiliency. A handful of more formal techniques have been studied in relation to telomere maintenance (telomerase or telomere length). Some of these are studies that have compared people cross-sectionally. For example, people practicing Zen meditation,[37] or loving-kindness meditation,[38] have longer telomeres than nonmeditators. But we don't know if a third factor (a "confound") might be causing that effect: Meditators have different values and behaviors. They may eat more kale chips, and fewer potato chips, than nonmeditators. The highest type of scientific evidence is controlled trials, where people are randomized to an active treatment or control group. You already heard about the meditators on a three-month mountain retreat. Good news: There have been more controlled trials that show that you don't have to leave home. A range of mind-body activities—Mindfulness Based Stress Reduction, yogic meditation, Qigong, and intensive lifestyle change—all promote better telomere maintenance. We describe these research studies in the "Master Tips" section at the end of Part Two (page 153).

TELOMERE TIPS

- Getting to know our habits of thinking is an important step toward wellbeing. Negative styles of thinking (hostility, pessimism, thought suppression, rumination) are common but cause us unnecessary suffering. Fortunately, they can be tempered.

- Increasing our stress resiliency—through purpose in life, optimism, unitasking, mindfulness, and self-compassion—combats negative thinking and excessive stress reactivity.

- Telomeres tend to be shorter with negative thinking. But they may be stabilized or even lengthened by practicing habits that promote stress resiliency.

RENEWAL LAB

TAKE A SELF-COMPASSION BREAK

Whenever you run into a situation that is difficult or stressful, try taking a self-compassion break. Kristin Neff, a psychologist at the University of Texas at Austin, has conducted extensive research on self compassion. Her early trials suggest that practicing self compassion can reduce rumination and avoidance, and increase optimism and mindfulness.[39] Here is a modified description of how to do it:[40]

Instructions: Recall a situation in your life that is bothering you, such as a health issue, a relationship conflict, or perhaps a work problem.

1. Say whatever word or expression that feels true to your situation:

"This is painful." "This is stressful." "This is really hard right now."

2. Acknowledge the reality of suffering: "Suffering is a part of life." Say something that reminds you of our common humanity and that this pain is not unique to you: "I'm not alone." "Everyone feels this way sometimes." "We all struggle in our lives." "This is part of being human."

3. Put your hands over your heart, or any other place that feels soothing and comforting, perhaps your belly or gently over your eyes. Take a deep breath, and say to yourself, "May I be kind to myself."

You can use a different statement that reflects your needs in the moment, including any of the following:

I accept myself as I am, a work in progress.
May I learn to accept myself as I am.
May I forgive myself.
May I be strong.
I will be as kind to myself as possible.

The first few times you take a self-compassion break, you may feel awkward, you may feel only a slight relief of pain. Keep at it anyway. When you feel pain, acknowledge it; remind yourself that you are not alone in your suffering; and put your hand on your heart, with kindness. Eventually, you will become proficient at giving yourself compassion, and you will find that these mini-breaks restore your resilient thinking.

MANAGE YOUR EAGER ASSISTANT

Most of us have been told at one time or another to beware the internal critic, that inner voice that whispers dark words into your psyche, telling you that you're not good enough, that everyone is against you, that you're thinking the wrong way. But that's counterproductive. The inner critic is part of you; get angry at it, and you're getting angry at yourself. Ultimately, you're just getting trapped in more negative thought patterns and causing yourself more discomfort.

Instead of fighting the critic, or trying to banish the critic, try accepting the critic. You can do this by thinking of the internal voice in friendlier terms. Darrah Westrup is a clinical psychologist and the author of several books about ACT, a therapy that's based on accepting life—and your mind—as it is. She suggests that you think of that voice in your head as an eager assistant. Your eager assistant isn't evil or cruel. You don't need to fire her or scold her or send her to the basement filing room. Your eager assistant is like a bright-eyed young intern, one who desperately wants to prove her worth by providing you with a steady stream of *well-intentioned but often misguided advice.*

It's unlikely you will ever get the eager assistant to stop offering a barrage of suggestions and comments about what you are doing, could have done better, or should do in the future. But you can manage your eager assistant. Be aware of her. Understand that what she's saying isn't necessarily "truth." Treat her in the same way you might manage an overly "helpful" young staff member at the office: Smile, nod, and tell yourself, "Oh, there goes my eager assistant again. She means well, but doesn't know what she's talking about here." That way, you're not in a battle with your own thoughts. By letting them be, they'll have much less influence over you.

WHAT'S ON YOUR TOMBSTONE?

The study of meditators in the Colorado Rockies found that a strong purpose in life appears to increase telomerase. Mindfulness meditation can increase your sense of purpose, but so can other activities. The following exercise may sound a bit ghoulish, but it can be clarifying:

Instructions: Write down the epitaph you'd like to see on your tombstone, the few words that you'd like the world to remember you by. To get ideas flowing, first ask yourself, what are you deeply passionate about? Here are examples we've heard:

- "Devoted father and husband."
- "Patron of the arts."
- "A friend to everyone."
- "Always learning, always growing."
- "An inspiration to all."
- "No one spread more love in one lifetime."
- "We make a living by what we get but we make a life by what we give."
- "If you don't climb the mountain, you can't view the plain."

There's not a lot of room on a tombstone! That's the whole point of the exercise; it forces you to articulate the one or two principles that are the most important to you. After doing this exercise, some people realize that they've been distracted by things that aren't all that important to them, and that it's time to attend to the priorities at the top of their list. Other people begin the exercise believing that they have a somewhat humdrum existence—but when they write their epitaph, they realize, joyfully, that they have been living in accordance with their highest goals.

SEEKING STRESS? YES, POSITIVE STRESS!

Is there something in your life that makes you nervous or excited? Is daily life too filled with predictable routine and not enough novelty to stretch your problem-solving, creative, or socializing skills? Maybe you could add more "challenge stress" to enliven your day. Doing cognitive exercises like crossword puzzles may be a good thing for maintaining your mental sharpness,[41] but they don't do much for living with vitality and purpose. You might consider stepping outside the box of daily routine and adding a new activity that is meaningful, fulfilling, and...antiaging. And as we saw with the Experience Corps, positive stress may even improve brain aging.

To pursue a new dream, we may need to stretch and get out of our comfort zone. New situations may make us anxious, but if we avoid them, we are missing opportunities to grow and thrive. Positive stress for you may be doing something you've wanted to try but were apprehensive about.

    Instructions: If you say yes to positive stress, close your eyes and think of what is at the top of your list. Take some time to think of something both exciting and feasible, a mini-adventure. Choose a small step toward that goal, something you can look into *today*. Bolster yourself with affirmations of your values and reappraisals to remind yourself that challenge stress is good stress.

ASSESSMENT: How Does Your Personality Influence Your Stress Responses?

Some personality traits can lead to bigger stress responses. To determine whether your personality could affect how your mind responds when stress comes your way, take the assessment on the next page. Whatever you learn about your personality, celebrate it. Personality is the spice of life, and knowledge about it is power. There is no right or wrong way to be. The point is to know yourself and be aware of your tendencies, not to change your personality. In fact, personality cannot change easily. It tends to be stable. Both genetics and life experiences have shaped our temperament. The more we are aware of our general tendencies, the more we can notice and live better with our natural habits of reacting to stress. And that can help us improve our telomere health.

A note to the skeptical: Some magazines or books contain personality assessments that are made up. They're fun, but they're not necessarily accurate. The personality assessments here include the actual measures used in research, reprinted with permission. (The hostility questions are an exception because those questions aren't available for public use. We've done our best to write our own questions that we feel will give you a good sense of your hostility level.) They're validated, meaning that they have been tested to see if they really measure the personality trait in question. (Note: These are shorter versions, but longer versions, those that include more questions, are more reliable.)

Instructions: For each question, circle the number that best describes how much you agree or disagree with the statement. As you take the assessments, pay attention to the words rather than the numbers. There are no right or wrong answers. Be as honest as you can.

WHAT'S YOUR THINKING STYLE?

How Pessimistic Are You?

1. I hardly ever expect things to go my way.	**4** Strongly Agree	**3** Agree	**2** Neutral	**1** Disagree	**0** Strongly Disagree
2. I rarely count on good things happening to me.	**4** Strongly Agree	**3** Agree	**2** Neutral	**1** Disagree	**0** Strongly Disagree
3. If something can go wrong for me, it will.	**4** Strongly Agree	**3** Agree	**2** Neutral	**1** Disagree	**0** Strongly Disagree
TOTAL SCORE					

Now calculate your total score by adding up the numbers you circled.

- If you scored between 0 and 3, you are **low** in pessimism.
- If you scored between 4 and 5, you are **average** in pessimism.
- If you scored 6 or above, you are **high** in pessimism.

How Optimistic Are You?

1. In uncertain times, I usually expect the best.	**4** Strongly Agree	**3** Agree	**2** Neutral	**1** Disagree	**0** Strongly Disagree
2. I'm always optimistic about my future.	**4** Strongly Agree	**3** Agree	**2** Neutral	**1** Disagree	**0** Strongly Disagree

3. Overall, I expect more good things to happen to me than bad.	**4** Strongly Agree	**3** Agree	**2** Neutral	**1** Disagree	**0** Strongly Disagree
TOTAL SCORE					

Now calculate your total score by adding up the numbers you circled.

- If you scored between 0 and 7, you are **low** in optimism.
- If you scored 8, you are **average** in optimism.
- If you scored 9 and above, you are **high** in optimism.

How Hostile Are You?

1. I usually know more than people I have to listen to or follow.	**4** Strongly Agree	**3** Agree	**2** Neutral	**1** Disagree	**0** Strongly Disagree
2. Most people cannot be trusted.	**4** Strongly Agree	**3** Agree	**2** Neutral	**1** Disagree	**0** Strongly Disagree
3. I am easily annoyed or irritated by other people's habits.	**4** Strongly Agree	**3** Agree	**2** Neutral	**1** Disagree	**0** Strongly Disagree
4. I get angry at other people easily.	**4** Strongly Agree	**3** Agree	**2** Neutral	**1** Disagree	**0** Strongly Disagree
5. I can be harsh or rough to people who are disrespectful or annoying.	**4** Strongly Agree	**3** Agree	**2** Neutral	**1** Disagree	**0** Strongly Disagree
TOTAL SCORE					

Now calculate your total score by adding up the numbers you circled.

- If you scored between 0 and 7, you are **low** in hostility.
- If you scored between 8 and 17, you are **average** in hostility.
- If you scored 18 and above, you are **high** in hostility.

How Much Do You Ruminate?

1. My attention is often focused on aspects of myself I wish I'd stop thinking about.	**4** Strongly Agree	**3** Agree	**2** Neutral	**1** Disagree	**0** Strongly Disagree
2. Sometimes it is hard for me to shut off thoughts about myself.	**4** Strongly Agree	**3** Agree	**2** Neutral	**1** Disagree	**0** Strongly Disagree
3. I tend to ruminate or dwell on things that happen to me for a really long time afterward.	**4** Strongly Agree	**3** Agree	**2** Neutral	**1** Disagree	**0** Strongly Disagree
4. I don't waste time rethinking things that are over and done with.	**0** Strongly Agree	**1** Agree	**2** Neutral	**3** Disagree	**4** Strongly Disagree
5. I never ruminate or dwell on thoughts about myself for very long.	**0** Strongly Agree	**1** Agree	**2** Neutral	**3** Disagree	**4** Strongly Disagree
6. It is hard for me to put unwanted thoughts out of my mind.	**4** Strongly Agree	**3** Agree	**2** Neutral	**1** Disagree	**0** Strongly Disagree
7. I often reflect on episodes in my life that I should no longer concern myself with.	**4** Strongly Agree	**3** Agree	**2** Neutral	**1** Disagree	**0** Strongly Disagree
8. I spend a great deal of time thinking back over my embarrassing or disappointing moments.	**4** Strongly Agree	**3** Agree	**2** Neutral	**1** Disagree	**0** Strongly Disagree
TOTAL SCORE					

Now calculate your total score by adding up the numbers you circled (be extra careful when adding your scores for questions 4 and 5—the numbers are reversed).

- If you scored between 0 and 24, you are **low** in rumination.
- If you scored between 25 and 29, you are **average** in rumination.
- If you scored 30 and above, you are **high** in rumination.

How Conscientious Are You?

I see myself as someone who . . .

1. Does a thorough job.	**4** Strongly Agree	**3** Agree	**2** Neutral	**1** Disagree	**0** Strongly Disagree
2. Can be somewhat careless.	**0** Strongly Agree	**1** Agree	**2** Neutral	**3** Disagree	**4** Strongly Disagree
3. Is a reliable worker.	**4** Strongly Agree	**3** Agree	**2** Neutral	**1** Disagree	**0** Strongly Disagree
4. Tends to be disorganized.	**0** Strongly Agree	**1** Agree	**2** Neutral	**3** Disagree	**4** Strongly Disagree
5. Tends to be lazy.	**0** Strongly Agree	**1** Agree	**2** Neutral	**3** Disagree	**4** Strongly Disagree
6. Perseveres until the task is finished.	**4** Strongly Agree	**3** Agree	**2** Neutral	**1** Disagree	**0** Strongly Disagree
7. Does things efficiently.	**4** Strongly Agree	**3** Agree	**2** Neutral	**1** Disagree	**0** Strongly Disagree
8. Makes plans and follows through with them.	**4** Strongly Agree	**3** Agree	**2** Neutral	**1** Disagree	**0** Strongly Disagree
9. Is easily distracted.	**0** Strongly Agree	**1** Agree	**2** Neutral	**3** Disagree	**4** Strongly Disagree
TOTAL SCORE					

Now calculate your total score by adding up the numbers you circled (be extra careful when adding your scores for questions 2, 4, 5, and 9—the numbers are reversed).

- If you scored between 0 and 28, you are **low** in conscientiousness.
- If you scored between 29 and 34, you are **average** in conscientiousness.
- If you scored 35 and above, you are **high** in conscientiousness.

How Much Purpose in Life Do You Feel?

1. There is not enough purpose in my life.	**0** Strongly Agree	**1** Agree	**2** Neutral	**3** Disagree	**4** Strongly Disagree
2. To me, the things I do are all worthwhile.	**4** Strongly Agree	**3** Agree	**2** Neutral	**1** Disagree	**0** Strongly Disagree
3. Most of what I do seems trivial and unimportant to me.	**0** Strongly Agree	**1** Agree	**2** Neutral	**3** Disagree	**4** Strongly Disagree
4. I value my activities a lot.	**4** Strongly Agree	**3** Agree	**2** Neutral	**1** Disagree	**0** Strongly Disagree
5. I don't care very much about the things I do.	**0** Strongly Agree	**1** Agree	**2** Neutral	**3** Disagree	**4** Strongly Disagree
6. I have lots of reasons for living.	**4** Strongly Agree	**3** Agree	**2** Neutral	**1** Disagree	**0** Strongly Disagree
TOTAL SCORE					

Now calculate your total score by adding up the numbers you circled (be extra careful when adding your scores for questions 1, 3, and 5—the numbers are reversed).

- If you scored between 0 and 16, you are **low** in life purpose.

- If you scored between 17 and 20, you are **average** in life purpose.
- If you scored 21 and above, you are **high** in life purpose.

SELF-ASSESSMENT SCORING AND INTERPRETATION

This assessment is simply meant to raise your awareness of your personal style. It is not meant to diagnose you or make you feel bad about being a certain way. Self-awareness of tendencies that make us more vulnerable to stress reactivity (and possibly telomere shortening, in several studies) is valuable! Awareness can help us notice unhealthy thought patterns and choose different responses. It can also help us know and accept our tendencies. As Aristotle reportedly said, "Knowing yourself is the beginning of all wisdom."

Dimensions that make us more vulnerable to stress	Score (Circle)		
Pessimism	High	Medium	Low
Hostility	High	Medium	Low
Rumination	High	Medium	Low

Dimensions that may help us be more stress resilient	Score (Circle)		
Optimism	High	Medium	Low
Conscientiousness	High	Medium	Low
Purpose in Life	High	Medium	Low

HOW DID WE DECIDE WHAT DETERMINES HIGH OR LOW SCORES?

In general, we determined the high, medium, and low score categories by looking at the data from large representative samples of people who had taken the test. We divided the population into thirds based on their scores. If you are in the top third (33 percent) of scores, you scored "high." If you are in the bottom third (33 percent) of scores, you scored "low." If you are in the middle, you scored "average." The actual studies used are described below.

The cutoff points should not be taken too literally. First of all, the comparisons are made to some large samples, but any one sample is never representative of everyone. There are always differences in how people score based on their race/ethnicity, sex, culture, and even age, that we could not take into account. Second, we assumed that there is a statistically "normal distribution" for scores for each measure, which means the same number of people score high as low, in the same symmetrical distribution pattern. In reality, few measures have scores that are perfectly normally distributed. Therefore, our cutoff points are not statistically perfect, nor are they perfectly accurate when applied to individuals.

THE PERSONALITY TYPES AND SCALES USED IN THIS ASSESSMENT

Optimism/Pessimism

Optimism is the tendency to expect or anticipate positive events and outcomes rather than negative ones. Optimism is characterized by a sense of hope and positivity about the future. **Pessimism** is the tendency to expect or anticipate negative events and outcomes rather than positive ones. Pessimism is characterized by a lack of hope and positivity about the future.

We used the "Life Orientation Test—Revised" (LOT-R) developed by Professors Charles Carver and Michael Scheier.[1] Optimism and pessimism are strongly related, but not totally overlapping, which means they are different aspects of personality. So it is helpful to examine them separately.[2] Two studies assessed the relationship with telomere length, and both found correlations with pessimism but not optimism.[3] That is not to say that optimism doesn't matter for health. It absolutely does, especially for mental health. It's just that with stress-related health outcomes, negative traits are often stronger predictors than positive traits, and they are more directly tied to stress physiology. Positive traits can buffer you from stress, and are weakly related to positive restorative physiology.

For scoring, we used the average levels on each LOT-R subscale from a study that tested over two thousand men and women who varied in age, gender, race, ethnicity, level of education, and socioeconomic class.[4]

Hostility

Hostility is thought to have cognitive, emotional, and behavioral manifestations.[5] The cognitive component, possibly the most important part of hostility, is characterized by negative attitudes toward others, colored by cynicism and mistrust. The emotional component ranges from irritation to anger to rage. The behavioral component is the tendency to act out verbally or physically in ways that could hurt others.

Hostility scales are not free to the public, so this scale includes items we created that should roughly measure hostility in the same way as the standardized research scales, particularly the most common one, the Cook-Medley Hostility Questionnaire, which is part of the MMPI Personality measure. We estimated the cutoffs based on the mean scores from a study of men from the Whitehall study, which used a short version of the Cook-Medley Hostility Questionnaire. This study found high hostility is related to shorter telomeres in men.[6]

Rumination

Rumination is "self-attentiveness motivated by perceived threats, losses, or injustices to the self."[7] In other words, rumination is the act of spending a significant amount of time thinking about and perseverating on past negative events in one's life and one's role.

We used the eight-item rumination subscale from the "Rumination-Reflection Questionnaire," developed by Professor Paul Trapnell.[8] To determine cutoffs, we used the item mean for the eight-item version.[9] While no studies have directly linked rumination to telomere length, we think it is an important part of the stress process. That's because it keeps stress alive in the mind and body long after the event has passed. In our daily diary study of caregivers, we have found that daily rumination is associated with lower telomerase.

Conscientiousness

Conscientiousness is the measure of the degree to which a person is organized, how careful a person is in certain situations, and how disciplined he or she tends to be.

We used the conscientiousness subscale from the "Big Five Inventory" developed by Professors Oliver John and Sanjay Srivastava.[10] This scale was used in a study that found a positive correlation between higher conscientiousness and longer telomeres.[11] For scoring, we used means from a large study that examined conscientiousness scores across ages.[12]

Purpose in Life

Purpose in life is not a typical personality dimension, but rather how much we are aware of having some explicit purpose or goal for our life. It is something that can change based on life experiences and personal growth. An individual who scores high on the Purpose in Life scale is characterized as having a strong sense of meaning in life, having aims and engaging in activites he or she strongly values, or having an outlook that gives life meaning.[13]

We used the "Life Engagement Test," a six-item scale developed by Professor Michael Scheier and colleagues.[14] For scoring, we used normative data from a study of 545 older adults (adjusted to a 0-to-3 scale).[15] No studies have directly linked purpose in life to telomere length. However, in a meditation retreat study, increased purpose in life was associated with higher telomerase. As reviewed in the previous chapter, purpose in life is linked to better health behaviors, physiological health, and stress resiliency.

CHAPTER SIX

When Blue Turns to Gray: Depression and Anxiety

Clinical depression and anxiety are linked to shorter telomeres—and the more severe these disorders are, the shorter your telomeres. These extreme emotional states have an effect on your cells' aging machinery: telomeres, mitochondria, and inflammatory processes.

Dave had been suffering from a viral infection—sneezing, cough, stuffy nose—for several days when suddenly he had trouble breathing. At first it was uncomfortable to draw a deep breath, and then it was agonizing. "I'm hyperventilating," Dave thought, and tried breathing into a paper bag. When that didn't help, he called his wife at work, and she agreed to pick him up at the street corner and drive him to an urgent care clinic. As he walked outside, the landscape seemed to darken, even though the day was bright. It was as if a deep shadow was overtaking his vision. His skin prickled. All this time, he continued to hyperventilate. When Dave arrived at the clinic, the nurses had to give him a mild sedative so that he could breathe well enough to describe his symptoms.

He was diagnosed with a panic attack, an intense episode of fear and anxiety. For Dave, the panic attack was actually a change from the depressive symptoms that have dogged him for most of his life. When he's depressed, he feels as if he has no possibilities, no future.

Every activity, even cracking an egg for a breakfast omelet, even just looking through his bedroom window, feels overwhelmingly strenuous and even physically painful. "I squint like I'm facing a strong wind," he says.

There are still people out there who don't take depression and anxiety seriously, who don't understand the phenomenal breadth and depth of suffering they cause. A global view helps put these problems into perspective: Mental disorders and substance abuse are the top causes of disability (defined as "productive days of life lost") worldwide, and the biggest player in this mix of disorders is depression, the "common cold" of psychiatry.[1] Heart disease, high blood pressure, and diabetes all develop earlier and faster in people with depression and anxiety. It's now much harder to write off depression and anxiety as "all in your head," because research has demonstrated that these states reach past your mind and soul, past your heart, past your bloodstream, and all the way into your cells.

ANXIETY, DEPRESSION, AND TELOMERES

Anxiety is characterized by excessive dread or worry about the future. It's not necessarily as dramatic as Dave's panic attack; often it's more like a steady, low-level thrum of unease. "I was standing at the edge of my driveway," said one woman we know, "waiting for my son to come home from a late hockey practice. I felt a little shaky, and my heart was beating fast. At first I thought that I was just worried about my son coming home safely. Then I realized that I felt this way most of the time. I finally asked myself, 'Is this normal?'" It isn't. The next week, she was diagnosed with generalized anxiety disorder.

Anxiety is a relatively recent subject of telomere research. People who are in the throes of clinical anxiety tend to have significantly shorter telomeres. The longer the anxiety persists, the shorter the telomeres. But when the anxiety is resolved and the person feels

139

better, telomeres eventually return to a normal length.[2] This is a grand argument in favor of identifying and treating anxiety. Sometimes, though, anxiety is hard to spot. As our friend came to realize, anxiety can seem normal when you're accustomed to feeling it, when it's the air that you breathe.

The depression–telomere connection has a more solid scientific literature behind it, possibly because depression is so pervasive: More than 350 million worldwide suffer from it. An impressive large-scale study of almost twelve thousand Chinese women by Na Cai and colleagues (at Oxford University and Chang Gung University in Taiwan) found that women who are depressed have shorter telomeres.[3] Depressed people, like anxious ones, show that dose-response we've talked about before. The more severe and prolonged the depression, the shorter the telomeres.[4] (See the bar chart in Figure 16.)

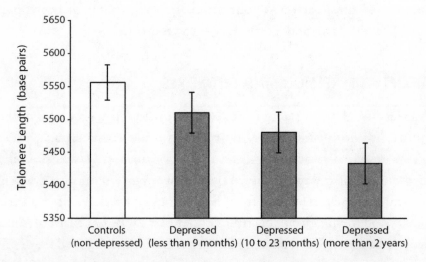

Figure 16: Duration of Depression Matters. The Netherlands Study of Depression and Anxiety has followed almost three thousand people, including those with depression and nondepressed controls. Josine Verhoeven and Brenda Penninx found those with current depression lasting less than ten months did not have significantly shorter telomeres than controls, but those with depression for more than ten months did.

A few studies suggest that short telomeres may directly lead to depression. People with depression have shorter telomeres in the hippocampus, an area of the brain that plays an important role in the disorder.[5] (They don't have shorter telomeres in other parts of the brain, just this part that is so crucial for mood.) Rats put under stress have less telomerase in the brain's hippocampus and less brain cell growth (neurogenesis), and they are more likely to develop depression.[6] However, when their telomerase is increased, the rats have more neurogenesis and do not become depressed. Cell aging in the brain may be one pathway to depression.

Here's an apparently odd phenomenon: depressed people have shorter telomeres but more telomerase in their immune cells. *What?* How could depression lead to shorter telomeres but more telomerase? This paradoxical combination appears in other situations, too: in people who are overloaded with stressful circumstances, people who haven't graduated from high school, men who are cynically hostile, and people who are at high risk for coronary disease. We believe that in these situations, the cells are producing more telomerase in response to telomere shortening, in a (sadly) ineffectual effort to build back the telomere segments that are being lost.

More support for this idea: Our colleague Owen Wolkowitz, a psychiatrist at UCSF, has been examining how telomerase might help with depression. Give depressed people an antidepressant (an SSRI), and their high telomerase levels climb up even further! The more that telomerase goes up, the more likely it is that their depression will lift.[7] It is possible that the efforts of the immune cells to replenish their lost telomeres reflect what is happening in the brain, with neurons doing the same thing. There may be a rejuvenation of sorts, in which more effective telomerase action (as opposed to ineffectual attempts by telomerase to elongate telomeres) could be promoting neurogenesis, the birth of new brain cells.

TRAUMA, DEPRESSION, AND THE REVERSAL OF STRESS EFFECTS

So far, most psychiatric disorders studied have been linked to telomere shortness, as shown by a meta-analysis.[8] Part of this could be due to the underlying stress leading up to onset of the disorders or from having the disorders. One of the most hopeful messages from the neuroscience of stress is that there is tremendous potential for brain plasticity, especially for the reversal of stress effects. We can overcome the effects of severe stress with antidepressants, exercise and other healthy buffers, and the passage of time. Telomere maintenance also shows plasticity. For example, in humans and rats it looks like telomeres likely shorten a small amount at the time of a stressful event, but in most cases they can eventually repair themselves.[9] Researcher Josine Verhoeven has examined patterns of recovery over time in the large Netherlands Study of Depression and Anxiety (NESDA) cohort: Major events in the last five years are associated with telomere shortness, but events in the distant past, more than five years ago, are not.[10] Similarly, having a current anxiety disorder is associated with shorter telomere length, but having one in the past is not, a finding that suggests telomeres can recover when an episode of anxiety is over. And the more years since the episode, the longer the telomere length.[11] Depression, however, appears to have a stronger imprint than stressful events or anxiety, as often people with past depression still have shorter telomeres.[12]

Cai's large Chinese study found a pattern suggesting that telomeres tend to rebound in people with past trauma—unless the person develops a serious depression. Then the telomeres remain short. It's as if trauma plus depression is a weight too heavy to bear. The good news is that even though telomeres can carry scars of past severe trauma plus depression, they can also be stabilized, and possibly lengthened, through activities that help boost telomerase. Telomeres can recover, thanks to telomerase.

Inside the cell, mitochondria are another important target of stress damage. Can mitochondria recover from stress as well? Mitochondria are critical to aging but have only recently been studied in terms of mental health. Mitochondria are the energy plants of the cell. Give them fuel, in the form of food molecules, and they'll process it into nutrient-rich molecules that power the cell. Some cells, like nerve cells, have one or two mitochondria; other cells need many more to keep up with their energy needs. Muscle cells, for example, typically have thousands of mitochondria. When you're in certain states of physical stress—if you have diabetes or heart disease—mitochondria can malfunction and cells won't receive enough energy. This can affect brain function, because the neurons don't have enough energy to fire. Your muscles may be weaker. The liver, heart, and kidneys—all organs that consume mass quantities of energy—suffer. One way to tell if cells are under major stress is to examine their mitochondrial DNA copy numbers, which tell us how hard the body is working to produce additional mitochondria to supplement the ones that are weary and damaged. In the Chinese study, it appeared that the greater the childhood adversity or depression, the shorter the telomeres and the higher the mitochondrial DNA copy numbers.

If you take mice and do some not very nice things to them (such as hanging them by their tails or forcing them to swim), they will, quite naturally, be stressed. Like humans, mice who are under stress develop an excessive number of mitochondria. It appears that their mitochondria are faulty and not working efficiently. Their cells are thus working desperately to increase their energy supply, with limited success. As you may imagine, stressed-out mice with high mitochondrial DNA copy numbers are not superenergetic. Further, their telomeres are 30 percent shorter. But given a month to recover from the stress, their telomeres and mitochondrial DNA become normal again. There is no lingering sign of accelerated aging.[13]

Biology can be molded by experience, then remolded. Cells can renew themselves. In a mouse's life, time-limited adversity can be mostly erased. Fortunately this appears true for many types of adversity in humans, too.

PROTECTING YOURSELF FROM DEPRESSION AND ANXIETY

Mental health is not a luxury. If you want to protect your telomeres, you need to protect yourself from the effects of depression and anxiety. Some proclivity toward these disorders is partially influenced by genes. But that does not mean everything is out of your control.

Depression is a complicated illness that lives in the emotions, in the thoughts, and in the body, and it's beyond the scope of this book to describe depression (or anxiety) in full. But here is one beautifully clear idea that drives some successful treatments: Depression is partly a dysfunctional response to stress. Instead of simply feeling the stress, depressed people tend to cope by using some of the negative thought patterns we've talked about. They try to suppress the bad feelings so that they can't be deeply felt, or they keep their problems alive by ruminating about them over and over and over. They criticize themselves. They feel irritable and angry, not just at whatever circumstances have caused their sorrow and stress, but at the fact of feeling sorrow and stress.

As we've said, this is a set of dysfunctional responses. Completely understandable, but dysfunctional nevertheless. Over time, this cycle can pull a person down past stress and into depression. Negative thoughts are like microtoxins—relatively harmless when your exposure is low, but in high quantities they are poisonous to your mind. Negative thoughts are not signs that you are truly unworthy or a failure. They are the substance of depression itself.

These counterproductive mental reactions are part of anxiety, too. Imagine this: You're at a cocktail party and you accidentally

144

call the hostess by the wrong name. She gasps a little, recovers with a tight smile, and corrects your error. You're embarrassed. Who wouldn't be? But for most of us, it's a pretty mild stress. Our cheeks may flush a little, we apologize, and we move on. But some people have what's known as anxiety sensitivity, and their bodies will produce an outsized physical response to the same event. Put these folks at a party, and if they make a mistake like forgetting a name, their hearts will race, they'll feel light-headed, and they may even think they're going to have a heart attack. It's a really uncomfortable state. A person with anxiety sensitivity naturally thinks, "Well, that was awful. From now on, I'll just avoid parties."

The problem with avoiding whatever is making you anxious is that the avoidance actually perpetuates the feelings of anxiety. You avoid the things you want and need to do, and you never learn that it's possible to tolerate the discomfort. In psychology terms, you never *habituate* to the stressful situation. Your life becomes smaller and smaller and tenser and tenser. Those anxious feelings bloom into a full-grown clinical disorder that interferes with your life. Just as depression is an intolerance of being sad, anxiety is an intolerance of feeling anxious. That's the reason that treatment for anxiety disorders often involves exposure to the triggers and sensations that make you the most anxious. You learn that you can ride through the waves of anxiety and survive it.

Stress plus this kind of avoidant coping style can lead to both anxiety and depression. Understanding how the mind works, why and how it gets stuck in these cycles of thoughts, is a key part of overcoming these disorders. If you have frequent painful feelings and thoughts that prevent you from living fully, it's important to protect your telomeres and seek help. Don't be one of the millions suffering untreated. Coping skills take a while to develop and embody as habits, so give yourself time to learn them with the help of a therapist, and don't give up.

WHERE YOU PLACE YOUR ATTENTION MATTERS

What if nothing is really wrong with you, except your thoughts that are insisting otherwise? When we're feeling sad, we naturally try to think our way out of it. We notice the gap between how we feel and how we want to feel. We start to live in that gap, wishing things could be different, trying hard to find an escape.

Mindfulness-based cognitive therapy, or MBCT, helps people out of that gap. It combines traditional strategies of cognitive therapy with mindfulness practices. Cognitive therapy helps you change distorted thoughts; mindfulness, as we've mentioned, helps you change how you relate to your thoughts in the first place. MBCT is potent against that great threat to your telomeres, major depression. It's been shown to be *as effective as an antidepressant*.[14] One of the bleakest aspects of depression is that it can become chronic; 80 percent of its sufferers experience a recurrence. John Teasdale, formerly of the University of Cambridge; Zindel Segal of the University of Toronto Scarborough; and Mark Williams of Oxford University have found that in people with three or more recurring depressions, MBCT slashes the risk of depression's return in half.[15] It's also becoming clear that MBCT helps with anxiety, and it's useful for any of us who struggle with difficult thoughts and emotions.

There are two basic modes of thinking, MBCT teaches. There is the "doing mode," which is what we do when we're trying to get out of the gap between how life is and how we want it to be. But there is another mode, and that is the "being mode." In the being mode, you can more easily control where you put your attention. Instead of frantically striving to change things, you can choose to do little things that bring you pleasure, and things that help you feel masterful and in control. Because "being" also allows you to pay more attention to people, you can more fully connect with them—a state that typically brings humans the most joy and contentment. Have you ever experienced the contentment of focusing all your attention on a small task, such as cleaning out a messy drawer? That's what being mode feels like.

Doing Mode versus Being Mode[16]

	Doing Mode (Automatic)	Being Mode
Where is your attention?	**Not noticing what you are doing**	**Paying attention to the moment**
What time period are you living in?	**Past or future**	**Now**
What are you thinking about?	**Absorbed in stressful ideas** Thinking about where I wish I was, not where I am right now. Nothing feels satisfactory.	**Absorbed in current experience** Able to fully taste, smell, touch, and feel. Able to fully connect with others. Radical acceptance of self, unconditional kindness.
Level of metacognition (thoughts about thoughts)	**Believe thoughts are true** Cannot observe the mind's workings. Mood is controlled by thoughts.	**Freedom from believing the thoughts** Understand the transient nature of thoughts, can observe thoughts as they come and go. Can tolerate unpleasantness.

This chapter may have been a bit disturbing. Many of us have suffered from one of these common maladies of the mind or know someone in our circles who has. But the larger story is that telomeres can recover from episodes of adversity and depression, and when they don't, they can still be protected going forward. You can fortify your resources to prepare for the next challenge that appears. You can adopt a resilient mind-set to allow for more peace in mind and body, as we've reviewed earlier, such as building an awareness of both your stress response style and your habits of thinking. You might also adopt the breathing break or heart-focused meditations at the end of this chapter.

The telomere scars we may carry from adversity are also evidence of a state that we might call "worn wise." Dealing with adversity can leave us wiser and stronger. One of my (Elissa's) favorite scales measures how much one has grown from trauma in various ways (feeling closer relationships, feeling more self-reliant, increasing faith or spirituality). We used this scale in our first study of caregivers. We were confused at first when we saw that the caregivers who had shorter telomeres also experienced more psychological growth. A closer look at this pattern revealed what was going on—it was all about duration of struggle. Those who had been providing care the longest had more telomere wear and tear but also had more life-enriching changes.[17] As Elisabeth Kübler-Ross, a Swiss psychiatrist who studied grief and mourning, once said, "The most beautiful people we have known are those who have known defeat, known suffering, known struggle, known loss, and have found their way out of the depths. These persons have an appreciation, a sensitivity, and an understanding of life that fills them with compassion, gentleness, and a deep loving concern. Beautiful people do not just happen."

TELOMERE TIPS

- Major stress, depression, and anxiety are linked to shorter telomeres in a dose-response fashion. But in most cases, these personal histories can be erased, thank goodness. For example, major events have no residue five years later.
- Mitochondria function is also impacted by severe stress and depression, but at least in mice there is recovery with time.
- The cognitive machinery that drives depression and anxiety includes exaggerated forms of negative thinking—intolerance of negative feelings and excessive avoidance that doesn't really work. Depression is characterized by being stuck in the "doing mode" mind-set, including ruminative thoughts, which create a vicious cycle.
- Mindfulness-based interventions can help move us from the common overdoing mode to a being mode and reduce rumination. See the "Three-Minute Breathing Break" in this chapter's Renewal Lab.

RENEWAL LAB

THREE-MINUTE BREATHING BREAK

The pioneers of MBCT (mindfulness-based cognitive therapy)—
John Teasdale, Mark Williams, and Zindel Segal—have developed
training programs to help people attain the being mode. It is best
to work with a practitioner to help you fully learn MBCT, but you
can easily take advantage of a core activity of MBCT, which is a
quick, three-minute "time in."[18] This breathing break is like practicing thought awareness. You might recognize that you are feeling something painful. You label your thoughts, allowing them to
exist in your mind, and know that they will pass. The lifetime of
an emotion, even a very unpleasant one, is no longer than ninety
seconds—unless you try to chase it away or engage with it. Then it
lasts longer. The breathing break is a way to keep negative emotions
from living past their natural life spans. You can make it a habit, so
it helps anchor you at any time, not just during hard moments. You
can picture this exercise like an hourglass—invite whatever is present in your mind broadly, then focus narrowly on the breath, and
then expand awareness out to your full surroundings. Here's our
modified version:

1. Becoming aware: Sit upright and close your eyes. Connect with your breathing for a long inhalation and exhalation. With this awareness, ask yourself, "What is my experience right now? What are my thoughts? Feelings? Bodily sensations?" Wait for the responses. Acknowledge your experience and label your feelings, even if they are unwanted. Notice any pushing away of your experience, and soften around it, allowing space for all that comes up in your awareness.

2. Gathering your attention: Gently direct your full attention to your breathing. Notice each inhalation and each long exhalation. Follow each breath, one after another. Use your breathing as an anchor into this present moment. Tune in to a state of stillness that is always there right below the surface of your thoughts. This stillness allows you to come from a place of being (versus doing).

3. Expanding your awareness: Sense your field of awareness expanding around you, around your breathing, around your whole body. Notice your posture, your hands, your toes, your facial muscles. Soften any tension. Befriend all of your sensations, greet them with kindness. With this expanded awareness connect with your whole being, encompassing all that is you in the present moment.[18]

This breathing break calms your body and offers you more control over your stress reactions. It shifts your thinking away from self-focus and the doing mode and moves it toward the peaceful being mode.

A HEART-FOCUSED MEDITATION: RELEASE MENTAL PRESSURE, RELEASE YOUR BLOOD PRESSURE

Our breath is a window into knowing and regulating our mind-body. It is an important switch influencing the communication between brain and body. It's sometimes easier to change our breath to relax than to change our thoughts. When we breathe in, our heart rate goes up. When we exhale, our heart rate goes down. By having a longer exhalation than inhalation, we can slow our heart rate more, and we can also stimulate the vagus nerve. Breathing into our lower belly (abdominal breathing) stimulates the sensory pathways of the vagus nerve that go directly to our brain, which has an even more calming effect. Dr. Stephen Porges, an expert in understanding the vagus nerve, has shown why there is a strong link between the vagus nerve, breath, and feelings of social safety. Many mind-body techniques naturally stimulate the vagus nerve, sending our brain those critical safety signals.

Exercises that slow breathing, such as mantra meditation or paced breathing, are a reliable way to lower blood pressure.[19] You are slowing down your body's need to be aroused. You are turning up the volume on your vagus nerve activity, suppressing the sympathetic nervous system and slowing your heart rate even more. The vagus also turns on growth and restorative processes.

For some, focusing on the heart can be more peaceful than focusing on the breathing, and can still slow the breathing rate. The heart has such a complex and responsive nervous system that it is thought of as the "heart brain." Below we provide a script for a short heart-focused meditation. It also has some words from loving-kindness meditations in it. This has not been tested to examine any telomerase effect, but as you can see above, breathing is the basis of relaxation.

If you'd like, try this script now:

HEART-FOCUSED MEDITATION

Sit comfortably. Take some long and slow inhales and even longer exhales.

Continue to breathe in, and breathe out, repeating a calming word or hold a beautiful image each time you breathe out slowly. Notice the pause between the breaths.

Become aware of your thoughts: "Where are my thoughts right now?" Smile at each of them as they pass through your mind; then return to your exhalation word or image.

Place your hands (palms or fingers) on your heart. You might say to yourself "Ahhh" as you exhale. Let the burdens you are holding release and flow out of your body.

"May I be in peace."

"May my heart be filled with kindness."

"May I be a source of kindness for others."

Picture your heart radiating love. Picture a pet or person that you feel complete love for. Let that love radiate out toward others in your life.

Continue to breathe in, and breathe out slowly. Notice where you are holding in tension. As you exhale, let yourself feel enveloped in safety, warmth, and kindness. ♥

MASTER TIPS FOR RENEWAL:
Stress-Reducing Techniques Shown to Boost Telomere Maintenance

The mind-body techniques and practices here have been shown in at least one study to increase telomerase in immune cells or lengthen telomeres. These effects are healthy for everyone, but they are especially important if you have high stress levels. Mind-body techniques, including meditation, Qigong, tai chi, and yoga, have been shown in clinical trials to improve wellbeing and reduce inflammation.[1] Many types of meditation also promote the mental skills for metacognition, changing how we see and respond to stressful events. While a very small number of people have negative experiences from meditation, in general there are minimal negative side effects of these practices and an abundance of positive benefits. The evidence so far does not suggest superiority of one type of meditation mind–body practice over another for telomere health.

We provide brief instruction or further resources on several of the methods below on our website, telomereeffect.com

MEDITATION RETREATS

The benefits of meditation on mental and physical health have been widely covered. When practiced regularly, it can help soothe negative thought patterns, help you connect more deeply with other people, and in some cases increase your sense of life purpose. Emerging research suggests it may even help your telomeres grow.

Researcher Cliff Saron of the University of California, Davis, has been studying the effects of residential retreats on experienced meditators. He found higher telomerase at the end of a three-month Shamatha retreat compared to a control group, and especially if the meditators had developed more purpose in life. A new study he conducted with researcher Quinn Conklin found that after three weeks of an intensive residential insight meditation retreat, experienced meditators had longer telomeres in their white blood cells than when they began, whereas the control group showed little change.[2]

As part of a collaborative group, we had an opportunity to conduct a highly controlled exploratory meditation study where both the retreat and control group lived at a resort. We examined the biological effects of a weeklong mantra-like meditation retreat led by Deepak Chopra and his colleagues at the Chopra Center, in Carlsbad, California. Women who had never or rarely meditated were randomized to either being on vacation at the resort or being in the meditation retreat. We compared them to women who were regular meditators and had already signed up for the same retreat. We found that after the week, everyone felt fantastic, showing dramatic improvements on all of our scales of wellbeing, regardless of what group they were in. Gene-expression patterns showed large changes—reductions in inflammation and stress pathways. Since these psychological and gene-expression improvements occurred in all groups, we think of this as the powerful vacation effect of unplugging from daily demands and staying at a resort. There appeared to be a meditation effect as well: Telomerase increased, but only in the experienced meditators, a finding that was marginally significant. And some other telomere-protective genes seemed more active.[3] Such intriguing findings point to greater benefits to cell aging for those already trained, but clearly have to be replicated.

MINDFULNESS-BASED STRESS REDUCTION (MBSR)

Mindfulness-based stress reduction, or MBSR, is a program created by Jon Kabat-Zinn at the University of Massachusetts Medical School for people with no or little meditation experience. Since 1979, around twenty-two thousand people have taken this program, and its benefits, such as reducing stress and physical symptoms like pain, have been well established.[4] MBSR includes training in the nature of the mind, mindful breathing, a mindful body scan (in which you slowly move your attention from your toes to the top of your head), and yoga. Taking a class in a group is a unique live experience, but for those who don't have access to MBSR locally, the University of Massachussetts Medical School's Center for Mindfulness offers an online course (http://www.umassmed.edu/cfm /stress-reduction/mbsr-online/). Their website also has a registry of trained MBSR teachers globally so you can see if you are near one.

In one study, people practicing MBSR increased their telomerase by 17 percent in a three-month period compared to a control group.[5] In another study, distressed breast-cancer survivors in the control group lost telomere base pairs—whereas the ones assigned to a form of MBSR focused on cancer recovery were able to maintain telomere length. A third group, one that received therapy based on emotional expression and support (supportive expressive group therapy), also maintained telomere length, a finding that provides encouraging news that the benefits of stress reduction on cell aging work across many different types of practices, not just meditation.[6] MBSR is great for anyone who wants to reduce stress; it's an especially good match for people suffering from chronic physical pain.

YOGIC MEDITATION AND YOGA

There are many types of meditation from many different traditions. Kirtan Kriya is a more traditional form of meditation from yoga principles that involves chanting and tapping of the fingers (called yoga mudras). Helen Lavretsky and Michael Irwin of UCLA conducted a study of people who were caring for family members with dementia, most of whom had at least mild symptoms of depression. Our lab measured their telomerase. When the caregivers practiced Kirtan Kriya for twelve minutes a day for two months, they increased their telomerase by 43 percent and decreased their gene expression related to inflammation.[7] (A control group listened to relaxing music; their telomerase increased, too, but only by 3.7 percent.) They were also less depressed, and their cognitive abilities improved.[8]

Unlike mindfulness meditation, which can help you develop metacognition and tolerate negative emotions, Kirtan Kriya works by putting you into a state of deep concentration and producing a calm, integrated state of body and mind. Afterward, your mind feels sharper and refreshed, as if you've just awakened from a great night's sleep.

A brief description can be found here: http://alzheimerspreven tion.org/research/12-minute-memory-exercise.

You may be wondering about Hatha Yoga, the type we commonly think of as exercise. It is a moving meditation that integrates physical postures, breathing, and a present mental state. Yoga has not been studied yet in relation to telomeres, but there is a tremendous research literature now on many health benefits of yoga. (For full disclosure, it is Elissa's personal favorite, and so hard not to mention.) Yoga improves quality of life and mood for people across different types of illnesses,[9] reduces blood pressure, and possibly inflammation and lipids.[10] Yoga has recently been shown to increase spine bone density if practiced long term.[11]

QIGONG

Qigong is a series of flowing movements; with its emphasis on posture, breathing, and intention, it is a kind of moving meditation.

Qigong is part of the wellness program of ancient Chinese medicine, a practice that has been developed and refined for more than five thousand years. Much like Kriya Yoga, Qigong induces a state of concentration and relaxation by integrating the body and mind. It is supported by thousands of years of practice but also the best kind of scientific evidence—randomized, controlled trials. For example, Qigong reduces depression[12] and may improve diabetes.[13] In a trial of Qigong on cell aging, researchers examined people with chronic fatigue syndrome. They found that people who practiced Qigong for four months had significantly greater increases in telomerase, and reductions in fatigue, than people who were assigned to a wait list.[14] A teacher showed the volunteers how to do Qigong for the first month, and then they practiced on their own at home, thirty minutes a day.

I've (Elissa) learned Qigong from Roger Jahnke, a doctor of Oriental medicine and an expert in medical Qigong. He recommends the practice both to prevent illness and to address particular health problems. The exercises are easy for anyone to do and can provide a strong sense of placid calm and wellbeing within minutes. (See examples on our website.) Many people are sensitive to how the body changes during this meditative activity, and they can feel a tingling sensation in the tips of their fingers (called chi/Qi sensation). This is partly due to the now-well-understood mechanisms of the relaxation response, which involves the activation of the parasympathetic nervous system and the dilation of blood vessels, creating new blood flow. This feeling is attributed to something from Chinese medicine that we have no concept for in our Western knowledge: chi/Qi energy flow.

INTENSIVE LIFESTYLE CHANGE: STRESS REDUCTION, NUTRITION, EXERCISE, AND SOCIAL SUPPORT

Dean Ornish, MD, president of the nonprofit Preventive Medicine Research Institute and clinical professor of medicine at UCSF, was the first to show that intensive lifestyle changes can reverse the progression of coronary heart disease. His program integrates stress-management techniques with other lifestyle changes. He wanted to see how this program might affect cell aging, so he studied this in men with low-risk prostate cancer. The men ate a diet high in plants and low in fat; they walked for half an hour, six days a week; they attended weekly support group sessions. They also practiced stress management on their own, with gentle yoga stretches, breathing, and meditation. In a prior randomized, controlled trial, this program was shown to slow or stop the progression of early-stage prostate cancer. At the end of the three months, the men's telomerase had also increased. Further, those who had greater reductions in their distressing thoughts about prostate cancer had the larger increases in telomerase, suggesting that stress reduction contributed to the improvements seen.[15] He followed a subgroup of these men for five years, and the ones who adhered to the program had significantly lengthened telomeres by 10 percent. His program for reversing heart disease is one of the very few behavioral programs that is now paid for by Medicare and many health insurance companies. You can find a certified provider for the heart disease program: https://www .ornish.com/ornish-certified-site-directory/.

PART III

HELP YOUR BODY PROTECT ITS CELLS

ASSESSMENT: What's Your Telomere Trajectory? Protective and Risky Factors

Next we are going to focus on the body—activity, sleep, eating. But before reading further, you are probably wondering how your telomeres are doing, and how you can find out. We pause here for a mini-assessment. We have telomeres in every cell of our body, in our different tissues, organs, and blood. They are correlated loosely—if we have short telomeres in our blood, we tend to have short telomeres in other tissues. A few commercial labs offer tests that measure the length of telomeres in your blood, but for individuals the usefulness of this is limited (see "Information about Commercial Telomere Tests" on page 333 and our website for a discussion of blood measures). It is more useful to assess the factors that are known to protect or damage telomeres and then, with the results of the assessment in mind, try to shift aspects of our daily lives so that our telomeres are more protected. That leads us to the Telomere Trajectory Assessment.

TELOMERE TRAJECTORY ASSESSMENT

You can assess the personal wellbeing and lifestyle factors that we know are related to telomere length. This assessment takes around ten minutes to complete and will help you identify the main areas you may want to improve.

When possible, we reprinted the actual scales used in the research described in this book. Research details for each scale are described after each section.

You will be asked about these areas:

Your Wellbeing

- current major stressful exposures
- clinical levels of emotional distress (depression or anxiety)
- social support

Your Lifestyle

- exercise and sleep
- nutrition
- chemical exposures

Do You Have Any Severe Stress Exposures?

Enter *1* next to any questions that apply to you and *0* next to any questions that don't apply. The situations must have been ongoing for at least several months for a score of 1.

Are you experiencing severe ongoing job stress, where you feel emotionally exhausted, burned out, cynical about your work, and fatigued, even when you wake up?	
Are you serving as a full-time caregiver for an ill or disabled family member and feeling overwhelmed by it?	
Do you live in a dangerous neighborhood and regularly feel unsafe?	
Are you experiencing severe extreme stress almost every day due to some chronic situation or a recent traumatic event?	
TOTAL SCORE	

Calculate your TOTAL SCORE by adding up items 1–4: _____

Circle the telomere points below that relate to your score.

Severe Stress Exposure Score	Telomere points (Circle)
If you scored 0, you have **low risk**.	2
If you scored 1, you have **some risk**.	1
If you scored 2 or higher, you have **high risk**.	0

Explanation: This Severe Stress Exposure Checklist is not a standardized scale. Instead, it measures whether you are experiencing an extreme situation that has been linked to shorter telomeres. For example, work-related emotional exhaustion,[1] being a caregiver for a family member with dementia,[2] and regularly feeling unsafe where you live[3] are related to shorter telomeres in at least one study, after controlling for factors such as BMI, smoking, and age. Any severe event has the potential to contribute to telomere shortening, if it occurs over years. Exposure alone is not a determinant—your response is important, too, as we discuss in chapter 4. Last, having one situation may be manageable, but having more than one severe chronic situation is more likely to exhaust one's coping resources. Multiple severe chronic situations are categorized here as a higher risk.

Any Mood Disorders?

Have you been currently diagnosed with depression or an anxiety disorder (such as posttraumatic stress disorder or generalized anxiety)?

Circle the telomere points below that relate to your score:

Clinical Distress Score	Telomere Points (Circle)
If you do not have a diagnosable condition, you are at **low risk**.	2
If you have been diagnosed with a severe condition, you are at **high risk**.	0

Explanation: From various studies, it appears the symptoms of moderate distress alone are not related to shorter telomeres, but actual diagnoses—which means symptoms that are severe enough to interfere with your daily life—are related.[4]

How Much Social Support Do You Have?

Answer the following questions about social support you typically receive from significant others, family, friends, and community members.

	1	2	3	4	5
1. Is there someone available to give you **good advice** about a problem?	None of the time	A little of the time	Some of the time	Most of the time	All of the time
2. Is there someone available whom you can **count on to listen to you** when you need to talk?	None of the time	A little of the time	Some of the time	Most of the time	All of the time
3. Is there someone available to you who shows you **love and affection**?	None of the time	A little of the time	Some of the time	Most of the time	All of the time
4. Can you count on anyone to provide you with emotional support (**talking over problems** or **helping you make a difficult decision**)?	None of the time	A little of the time	Some of the time	Most of the time	All of the time
5. Do you have as much contact as you would like with someone you feel close to, someone in whom you can trust and confide in?	None of the time	A little of the time	Some of the time	Most of the time	All of the time
TOTAL SCORE					

Now calculate your total score by adding up the numbers you circled.

Circle the telomere points below that relate to your score.

Social Support Score	Telomere Points (Circle)
If you scored 24 or 25, you are **high** in social support.	2
If you scored between 19 and 23, you are **average** in social support.	1
If you scored between 5 and 18, you are **low** in social support.	0

Explanation: This questionnaire is the five-question version of the ENRICHD Social Support Inventory (ESSI), originally created to assess social support of post–heart attack patients and used in epidemiological studies.[5] Versions of this questionnaire have been used in studies relating telomere length to social support.[6]

Cutoffs for the social support categories are approximations from data from a large study, and the effects in this study were found in the oldest age group only.[7] The ENRICHD trial used the score of 18 as a lower cutoff point to define people who were low on social support.

How Much Physical Activity Do You Do?

During the past month, which statement best describes the kinds of physical activity you usually did?

1. I did not do much physical activity. I mostly did things like watching television, reading, playing cards, or playing computer games, and I took one or two walks.

2. Once or twice a week, I did **light activities** such as getting outdoors on the weekends for an easy walk or stroll.

3. About three times a week, I did **moderate activities** such as brisk walking, swimming, or riding a bike **for about 15–20 minutes each time**.

4. Almost daily (five or more times a week), I did **moderate activities** such as brisk walking, swimming, or riding a bike **for 30 minutes or more each time**.

5. About three times a week, I did **vigorous activities** such as running or riding hard on a bike **for 30 minutes or more each time**.

6. Almost daily (five or more times a week), I did **vigorous activities** such as running or riding hard on a bike **for 30 minutes or more each time**.

Circle the telomere points that relate to your score.

Exercise Score	Telomere Points (Circle)
If you chose options 4, 5, or 6, you are at **low risk**.	2
If you chose option 3, you are at **average risk**.	1
If you chose option 1 or 2, you are at **high risk**.	0

Explanation: This questionnaire is the Stanford Leisure-Time Activity Categorical Item (L-CAT) (permissions granted by Nature Publishing Group).[8] The L-CAT assesses six different levels of physical activity. Scores of 4, 5, or 6 meet the CDC recommendations for aerobic activity (150 minutes of moderate exercise, like brisk walking, or 75 minutes of vigorous exercise like jogging; note the CDC also recommends muscle-strengthening activities at least two days a week). As we explain in chapter 7 ("Training Your Telomeres"), if you are fit and do regular exercise, there doesn't appear to be an upper limit to its benefits as long as you don't overdo it during workouts and you give yourself recovery time after big workouts. Think "regular exerciser," not "weekend warrior."

People who are more physically active appear to be better buffered from the telomere shortening that occurs due to extreme stress than people who are less active.[9] Additionally, an intervention showed that exercising forty-five minutes three times a week led to increases in telomerase.[10]

What Are Your Sleep Patterns?

During the past month, how would you rate your sleep quality overall?	**0** Very good	**1** Fairly good	**2** Fairly bad	**3** Very bad
How many hours of sleep do you get on average each night (not including lying in bed awake)?	**0** 7 hours or more	**1** 6 hours	**2** 5 hours	**3** Less than 5 hours

Circle the telomere points that relate to your score.

Sleep Score	Telomere Points (Circle)
If you scored a 0 or 1 on both questions, you're at **low risk**.	2
If you scored a 2 or 3 on one question, you're at **some risk**.	1
If you scored 2 or 3 on both questions, or you have poorly treated sleep apnea, you have **high risk**.	0

Explanation: The item on sleep quality is from the Pittsburgh Sleep Quality Index (PSQI), which assesses quality and disturbances of sleep.[11] Several studies relating telomere length to sleep have used the PSQI to measure sleep quality.[12] Duration of sleep is also important. If you report sleeping at least six hours per night and describe your sleep as good or very good, you're at low risk. If you report either poor sleep quality or shorter sleep durations, this adds risk. And if you report both poor sleep quality and shorter sleep durations, this is categorized as high risk. Since studies have not tested for an additive effect of both short and poor sleep, we are making an assumption that having both is worse.

If you have sleep apnea and do not treat it nightly, you are also at risk.

What Are Your Nutrition Habits?

How often do you have the following? Circle *1* or *0* for each question.

1. Omega-3 supplements, seaweed, or fish that contains high omega-3 oils:	
3 servings or more a week of these products?	1
Less than 3 times a week	0
2. Fruits and vegetables:	
At least daily?	1
Not every day	0
3. Sugared sodas or sweetened beverages (not including when you add your own sugar to coffees or teas, which typically adds up to substantially less sugar than in commercially sweetened drinks):	
At least one 12-ounce drink on most days	0
Not regularly	1
4. Processed meat (sausage, lunch meats, hot dogs, ham, bacon, organ meats):	
Once a week or more	0
Less than once a week	1
5. How much of your diet is whole foods (whole grains, vegetables, eggs, unprocessed meats) versus processed food (packaged or processed with salts and preservatives)?	
Mostly eat whole foods	1
Mostly eat processed foods	0

Add your total points from all five nutrition questions, creating a score between 0 and 5.

TOTAL SCORE (sum of items 1–5): _____

Circle the telomere points that relate to your score.

Telomere Nutrition Score	Telomere Points (Circle)
If you scored a 4 or 5, you have excellent protection from diet.	2
If you scored a 2 or 3, you have **average risk** from diet.	1
If you scored 0 or 1, you have **high risk** from diet.	0

Explanation: The frequencies were extrapolated from telomere studies.

For omega-3s, food sources are best. If you rely on supplements, try algae-based products instead of fish for sustainability reasons. People with higher blood levels of omega-3 fatty acids (DHA, or docosahexaenoic acid, and EPA, or eicosapentaenoic acid) have slower attrition over time.[13] Those who ate a half serving of seaweed each day had longer telomeres later in life.[14] An omega-3 supplement study found that dose didn't matter as much as what's absorbed in your blood: Taking either a 1.25 gram or 2.5 gram omega-3 supplement decreased the ratio of omega-6s to omega-3s in the blood at least to some extent for most people, which in turn was associated with an increase in telomere length.[15] It's hard to know how much your body will absorb, but it should be sufficient to have fish several times a week, or take a gram of omega-3 oils a day.

While supplements are also associated with longer telomere length, real foods with antioxidants and vitamins are superior if available (i.e., lots of vegetables and some fruit).

Sugar carbonated beverages are linked to shorter telomeres in three studies,[16] and it is prudent to assume that daily consumption would be a sufficient dose to have an effect, as suggested in one of these studies. Most sweetened beverages have over 10 grams of sugar, typically 20 to 40 grams.

For processed meat, one study showed that those in the highest quartile of the sample—those who ate processed meats once a week (or a tiny portion each day)—had shorter telomeres.[17]

How Much Are You Exposed to Chemicals?

Circle either *Yes* or *No* for the following questions.

Do you regularly smoke cigarettes or cigars?	Yes	No
Do you do regular agricultural work with pesticides or herbicides?	Yes	No

Do you live in a city with very heavy traffic-related pollution?	Yes	No
Do you work in a job with heavy exposure to chemicals listed on the Telomere Toxins table (see page 266), such as hair dyes, household cleaners, lead or other heavy metal exposure (for example, in a car mechanic's shop)?	Yes	No

Telomere Chemical Exposure Score	Telomere Points (Circle)
If you answered all nos, you have **low risk** from chemical exposures.	2
If you answered yes to one or more, you have **high risk**.	0

Explanation: Here we listed exposures that have been linked to telomere shortness in at least one study. The exposures include smoking,[18] pesticide exposure,[19] chemical exposures in dyes and cleaners,[20] pollution,[21] lead exposure,[22] and exposures in a car mechanic shop.[23]

How Did You Score Overall?

Area	Telomere Points (Circle)		
WELLBEING:	high risk	average	low risk
Stress exposures	0	1	2
Clinical emotional distress	0	1	2
Social support	0	1	2
LIFESTYLE:			
Exercise	0	1	2
Sleep	0	1	2
Nutrition	0	1	2
Chemical exposures	0	1	2
Total Score (range 0 to 14) _____			

How to Understand Your Total Telomere Trajectory

The summary score is a way to show overall risk and protection of your telomere rate of decline. If you have a high score, you likely have great telomere maintenance. Keep up the good work! The most useful way to use this assessment is to focus on individual areas rather than your total score. **If you scored a 2 on any area in the summary grid, you are doing a great job at telomere protection. You are doing more than simply dodging risk. Typically, this score means you are performing protective behaviors every day, engaging in the daily work creating the foundation of a good healthspan.**

If you scored a 0 (high-risk category), you are likely to experience the typical age-related telomere decline, made worse by risk factors, but ones that you hopefully can gain more control over.

Choose an Area to Work On

The best way to use this chart is to notice the areas in which you scored 0, and then decide which will be the easiest for you to change. If you don't have any 0s, choose a category in which you scored a 1. Wherever you begin, **we suggest you choose only one area to work on at a time**. Make a commitment to improve one small thing in this area. Put a reminder of the change you're trying to make on your bedside table, or set a reminder alarm on your phone to go off at a helpful time of day. At the end of Part Three, you will see some tips to get you started on your new goal.

CHAPTER SEVEN

Training Your Telomeres: How Much Exercise Is Enough?

Exercise reduces oxidative stress and inflammation, so it's not surprising certain exercise programs have also been shown to increase telomerase. However, weekend warriors should beware: overdoing a workout can actually promote oxidative stress, and overdoing it on a chronic basis (overtraining) may cause serious damage to you and your telomeres.

In May 2013, Maggie ran her first ultramarathon. She had been a strong contender in shorter races, and she liked the idea of pushing herself to run very long distances, like this hundred-mile race through the desert. She didn't even let herself hope that she might take a top spot; she just wanted to finish. Halfway through the ultramarathon, one of Maggie's friends met up with her and said, "Did you know you're in thirteenth place? You could finish in the top ten!"

Maggie decided to dig deeper. Over the next several hours, she overtook the twelfth-place runner, and then the eleventh, and then the tenth. She crossed the finish line in tenth place, guaranteeing that she would be invited back the next year to run in a place of honor.

That summer, Maggie ran three more ultramarathons: another

hundred-miler in June, and two more in July and August. She felt great. In September, she decided that instead of taking a long recovery period from her grueling workout schedule, she'd train for another ultra in December. Then, suddenly, a few weeks into training, Maggie virtually stopped sleeping. She spent entire nights wide awake; she would sit up in bed and watch as her phone lit up in the morning, its alarm ringing. "I've never done drugs, but I imagine this is what being on speed is like," Maggie says. "I couldn't sleep, and I wasn't tired. I had a ton of energy. It was *very* weird."

Maggie continued to train. Then the illnesses set in: colds, the flu, other viruses. She tried cutting back on her workouts, but she didn't notice an improvement in her symptoms, so she resumed her schedule. Then, in early winter, her body broke down. She couldn't complete her workouts. She could barely get to work. She could barely get out of bed.

Maggie had nearly all the signs of overtraining syndrome, an unofficial diagnosis that is characterized by sleep changes, fatigue, moodiness, vulnerability to illness, and physical pain.

When Maggie reminisces about her "grand slam summer" of ultramarathons, the people around her have mixed reactions. Some are judgmental—they declare, almost gleefully, that such intense exercise is bound to be bad for the human body. Others feel guilty. Despite Maggie's troubles, they feel that something is wrong with them if they're not working out at this elite level. Others use Maggie's experience as an excuse not to work out at all.

Exercise can be a confusing topic; it can also be an emotional one. But telomeres offer some clarity. Telomeres do not need extreme fitness regimens to thrive, and that is good news for all of us who feel discouraged when we meet people like Maggie, who spent her grand slam summer taking her body to the breaking point and then past it. Another piece of good news is that telomeres appear to respond powerfully to many different levels and types of exercise. In this chapter, we'll show what the range of healthy exercise looks like and how you can gauge whether you are doing too little—or, as in Maggie's case, too much.

TWO PILLS

Let's pretend you're at a drugstore of the future. You consult with the pharmacist, who gives you a choice between two pills. You point to the first one and ask what it does.

The pharmacist ticks the benefits off on her fingers. "Lowers your blood pressure, stabilizes your insulin levels, improves your mood, increases your calorie burn, fights osteoporosis, and cuts your risk of stroke and heart disease. Unfortunately, its side effects include insomnia, skin rash, heart problems, nausea, gas, diarrhea, weight gain, and lots of others."

"Hmmm," you say. "How about the second pill? What does it do?"

"Oh, it's got the same benefits," the pharmacist says brightly.

"And the side effects?" you ask.

She beams. "There aren't any."

The first pill is imaginary, a fantasy synthesis of beta-blockers to control high blood pressure, statins to reduce cholesterol, diabetes drugs to regulate insulin, antidepressants, and osteoporosis medications.

The second pill is real, sort of. It's called exercise. People who exercise live longer and have a lowered risk of high blood pressure, stroke, cardiovascular disease, depression, diabetes, and metabolic syndrome. *And* they avoid dementia for longer.

If exercise is like a drug that pumps wondrous effects through your entire physiology, how does it work? You already know the macro view of exercise. It increases blood flow to your heart and your brain, builds muscle, and strengthens your bones. But if you could take a powerful microscope to exercise's effects, and peer into the heart of human cells when they are regularly exercised, what would you see?

Calmer, Slimmer, and Better at Fighting Free Radicals: The Cellular Benefits of Exercise

People who exercise spend less time in the toxic state known as oxidative stress. This noxious condition begins with a free radical, a molecule that is missing an electron. A free radical is rickety, unstable, incomplete. It craves the missing electron, so it swipes one from another molecule—which is now unstable itself and needs to steal a replacement electron of its own. Like a dark mood that is passed from one person to another, each person feeling a little better the bad feelings are dumped onto someone else, oxidative stress is a state that can shear through a cell's molecular population. It's associated with aging and onset of the diseasespan: cardiovascular disease, cancer, lung problems, arthritis, diabetes, macular degeneration, and neurodegenerative disorders.

Fortunately, our cells also contain antioxidants, which offer natural protection against oxidative stress. Antioxidants are molecules that can donate an electron to a free radical but still remain stable. When an antioxidant gives an electron to a free radical, the chain reaction ends. An antioxidant is like a wise friend who says, "Okay, tell me all your bad feelings; I'll listen and you'll feel better, but I'm not going to let you make me feel bad, too. And I'm definitely not going to pass your black mood on to someone else."

In an ideal situation, your cells have enough antioxidants to keep up with the need to neutralize free radicals in your body. Free radicals will never be completely eradicated from our bodies. They are continuously being made by the very processes of life—they occur normally through metabolism. In fact, very small numbers of free radicals are important for the normal communication processes in our cells. But radicals can also be created in excess when you're exposed to environmental stresses like radiation and smoking, or to severe depression. The danger seems to occur when free radicals build up. And when you have more free radicals than antioxidants, you enter an imbalanced state of oxidative stress.

That's one reason exercise is so valuable. In the short term, exercise actually causes an increase in free radicals. One reason is that you're taking in more oxygen. Most of those oxygen molecules are used to create energy from special chemical reactions in the mitochondria in your cells, but an unavoidable by-product of these vital processes is that some of them also form free radicals. But that short-term response creates a healthy counterresponse: The body steps up by producing more antioxidants. Just as short-term psychological stress can toughen you up and increase your ability to handle difficulty, the physical stress of moderate-intensity regular exercise ultimately improves the antioxidant–free radical balance so that your cells can stay healthier.

Your cells soak up the benefits of exercise in other ways, too. When you exercise regularly, the cells in your adrenal cortex (located inside your adrenal glands) release less cortisol, the notorious stress hormone. With less cortisol, you feel calmer. With regular exercise, cells throughout your body become more sensitive to insulin, which means your blood sugar is more stable. If you want to avoid the common midlife trifecta of stress, belly weight gain, and high blood sugar, you need to exercise.

Immunosenescence: Exercise Can Keep You in the Healthspan Longer

Immunosenescence is an important process underlying increased sickness and malignancy as we age. As a result of immunosenescence, you experience higher circulatory levels of proinflammatory cytokines, molecules that can spread inflammation through the body like a fire whipped along by gusts of wind. This hastens more of your T-cells toward senescence so they can't do their work of fighting off sickness. Some senescent immune cells, as you know from earlier in the book, can even go rogue. These aging immune cells can leave you more vulnerable to the kind of nasty bugs that can put you into a hospital bed. If you have a lot of immunosenescent cells and you get a vaccine for pneumonia or this year's strains of flu, there's a good chance the vaccine won't "take" and that you'll end up feverish and coughing anyway.[1] Your aging cells make it harder for you to enjoy the benefits of preventive medicine.

However. Compared to couch potatoes, people who exercise regularly have lower inflammatory cytokine levels, respond more successfully to vaccinations, and enjoy a more robust immune system. Immunosenescence is a natural process that happens with age...but people who exercise may be able to delay it until the end of life. As the exercise and immunology researcher Richard Simpson has said, these and other signs "indicate that habitual exercise is capable of regulating the immune system and delaying the onset of immunosenescence."[2] Consider exercise an excellent bet for keeping your immune system biologically young.

WHAT KIND OF EXERCISE IS BEST FOR TELOMERES?

Exercise helps protect your cells by warding off inflammation and immunosenescence. Now there is an additional explanation of exercise's cellular benefit: exercise helps you maintain your telomeres. This was true even in a study of 1,200 twin pairs, which allowed exercise's effects to be teased apart from effects of genetics: the active twin had longer telomeres than the less active twin.[3] After adjusting for age and other factors that affect telomeres, by statistically removing their effect, the relationship between telomeres and activity was laid bare. And it's not just that exercise is helpful; we also know that sedentariness itself is terrible for metabolic health. Now several studies have found that sedentary people have shorter telomeres than people who are even a little more active.[4]

But are all types of exercise equal when it comes to cell aging? Researchers Christian Werner and Ulrich Laufs of Saarland University Medical Center in Homburg, Germany, tested three types of exercise in a small but exciting study. Their results hint that exercise really may increase telomerase's replenishing action—and they help us understand which kinds of exercise are best for keeping our cells healthy. Two kinds of exercise stood out. Moderate aerobic endurance exercise, performed three times a week for forty-five minutes at a time, for six months, increased telomerase activity twofold. So did high-intensity interval training (HIIT), in which short bursts of heart-pounding activity are alternated with periods of recovery. Resistance exercise had no significant effect on telomerase activity (although it had other benefits; the researchers concluded that "resistance exercise should be complementary to endurance training rather than a substitute"). And all three forms of exercise led to improvements in telomere-associated proteins (such as telomere-protecting protein TRF2) and reduced an important marker of cellular aging known as p16.[5] They also found that regardless of exercise type, those who increased their aerobic fitness the most had greater increases in telomerase activity. This tells us it's the underlying cardiovascular fitness that matters most.

So try to do moderate cardiovascular exercise or HIT. Either is great. Our Renewal Lab at the end of this chapter will show you these evidence-based workouts for strengthening your telomeres. You may not want to restrict yourself to just one type of workout, though. We benefit from variety. In a study of thousands of Americans, the more categories of exercise—from walking to biking to strength training—that people engaged in, the longer their telomeres.[6] And this is a reason to perform strength training. Even though strength training doesn't appear to be significantly related to longer telomeres, it helps maintain or improve bone density, muscle mass, balance, and coordination—all of which are vital for aging well.

How, exactly, does exercise strengthen telomeres?

Maybe the wonderful cellular effects of exercise, including less inflammation and oxidative stress, are good for telomeres. Or maybe exercise is good for telomeres because it prevents stress from causing some of its usual damage. The stress response can leave cellular damage and debris in its wake—but exercise switches on autophagy, the housekeeping activity in the cell that eats up those damaged molecules and recycles them.

It's also possible that exercise improves telomeres directly. For example, getting on the treadmill induces an acute stress response, which increases the expression of TERT, a telomerase gene.[7] Athletes have higher expression of TERT than sedentary people.[8] Exercise releases a newly identified hormone, irisin, that boosts metabolism and in one study was associated with longer telomeres.[9]

But no matter how the exercise–telomere connection works, what's most significant is that exercise is essential for your telomeres. To keep your telomeres healthy, you need to work them out. For the workouts that have been shown to improve telomere maintenance, see the Renewal Lab.

EXERCISE AND INTRACELLULAR BENEFITS

Exercise leads to myriad beautiful intracellular changes. Exercise causes a brief stress response, which triggers an even bigger restorative response. Exercise damages molecules, and damaged molecules can cause inflammation. However, early on in a bout of exercise, exercise induces autophagy, the Pac-Man-like process that eats up damaged molecules. This prevents inflammation. Later in the same exercise session, when there are too many damaged molecules, and autophagy can no longer keep them under control, the cell dies a quick death (called apoptosis), in a cleaner way that doesn't lead to debris and inflammation.[10] Exercise also increases the number and quality of those energy-producing mitochondria. In this way, exercise can reduce the amount of oxidative stress.[11] After exercise, when your body is recovering, it is still cleaning up cell debris, making cells healthier and more robust than before exercise.

GAUGE YOUR TELOMERE FITNESS

It's not simply working out that is crucial for telomere health. As we hinted earlier, fitness, the ability to perform physical tasks, is important, too. It is very possible for someone to perform light regular exercise, but not be fit. And some lucky people can be fit without exercising, especially when they're young. (Think of twentysomethings who can successfully complete a long, arduous hike even if they haven't worked out since high school.) For telomere health, you need to get regular exercise *and* you need to be fit.

But how fit do you need to be? Do you need to be capable of running ultramarathons, like Maggie? Swim five miles in the open water? Do you have to be like one of our Midwestern friends, who spends Saturday mornings in October at races featuring "zombies" who chase her through fields of corn? Our cultural standards for fitness are getting higher and higher, and it can be hard to know whether you are fit enough to stay healthy.

Fitness is, in fact, crucial for telomere health.[12] But you may be relieved to know that significant telomere benefits are gained by having a very moderate, achievable level of fitness. Our colleague Mary Whooley, of UCSF, put a group of adults, all of whom have heart disease, on a treadmill. They began by walking, with the incline and speed gradually increased until they could do no more. The results were clear: The less exercise capacity they had, the shorter their telomeres.[13] The people with the lowest cardiovascular fitness couldn't even sustain a brisk walk, whereas those most fit sustained the equivalent pace to taking a hike. Those with low fitness had fewer base pairs by an amount that translates to about four years of extra cell aging, compared to the fit group.

Are you able to mow your lawn? Shovel snow? Carry your own clubs while playing golf? If not, you are in this category of low fitness. There are easy ways to slowly and safely build your capacity. Check with your doctor and then consider our walking plan, in the Renewal Lab. On the other hand, if you can walk vigorously or maintain a light jog for forty-five minutes, three times a week, you are fit enough to support your telomere health. Remember that fitness and exercise are related, but separate. Even if you are naturally fit, you still need an exercise program to keep your telomeres healthy.

TOO MUCH EXERCISE?

Moderate exercise and fitness are clearly wonderful for telomeres, but what about Maggie, the ultramarathon runner? Are her telomeres longer because she took exercise to the extreme? Are they shorter? Few of us run ultramarathons, but as more people participate in endurance sports, questions like these become more urgent.

Most extreme exercisers can breathe a sigh of relief. One remarkable study of ultrarunners found that their cells were the equivalent of sixteen years younger than those of their sedentary counterparts.[14] Does this mean that we should all sign up for the next hundred-mile race? Not at all. Those ultrarunners were compared to *sedentary* people. When endurance athletes are compared to more ordinary runners, who might run around ten miles a week, you find that both groups have nice, healthy telomeres compared to the more sedentary group—and that there appears to be no additional benefit for the ultra–long-distance endurance group in terms of telomeres.[15]

Endurance athletes sometimes worry about whether it's safe to continue their extreme training year after year, as opposed to training for a single endurance event and then returning to a more typical exercise routine. One study looked at older men who had been elite athletes when they were younger. Their telomeres were similar in length to other men their age, so their many years of extremely vigorous training didn't appear to have had a cumulative wearing effect.[16] Another German study examined a group of senior "master athletes" who had been competing in endurance races since their youth. Most of them still compete, just at a slower rate (such as taking eight hours instead of two to complete a marathon). The long-term athletes both looked younger and had less telomere shortening compared to matched controls.[17] Another study examined years of exercise and found longer telomeres in people who had actively exercised for the past ten years or more.[18] It appears important to start exercising when you are young—but don't be discouraged. It's never too late to start, and benefits always await you.

Maggie, however, may be in some trouble. One study of extreme exercisers found that they had shorter muscle telomere length—but only if the exercisers suffered from a fatigue-overtraining syndrome.[19] When athletes develop a fatigue syndrome, as Maggie did, it is a sure sign they have overtrained and damaged their muscles to the extent that they can't be easily repaired. Progenitor cells (also known as satellite cells) repair muscle tissue that has been damaged—but it is thought that overtraining damages those crucial cells, leaving them less able to do their repair work. It appears to be overtraining, not extreme exercise, that is damaging to telomeres, at least in the muscle cells.

Overtraining is defined by too much training time relative to rest and recovery. It can happen to anyone, from beginning runners to professional athletes, and it happens when you don't support your body with enough rest, nutrition, and sleep. Psychological stress can contribute, too. Some warning signs of overtraining include fatigue, moodiness, irritability, trouble sleeping, and susceptibility to injuries and illness. The cure is rest—which sounds easy but is hard for athletes who are accustomed to pushing themselves.

Any discussion of overtraining is complicated, because there is no set point that constitutes "too much exercise." That point is different for everyone, and it depends on individual physiology and level of training. If telomeres tell us anything, they remind us of how context-dependent health can be. What is good for one person may be harmful to another. If you are an extreme athlete, make sure you're working closely with a trainer or physician so that any signs of overtraining can be spotted early.

In general, it's a good idea to begin *any* exercise program slowly, gradually working up to better fitness. Weekend warriors who sit in the office for five days and then overdo it on the weekend, pushing themselves to break down a lot of muscle at once, will feel fatigued and sometimes even nauseated. They are not doing their bodies any favors. Remember that exercise initially creates additional oxidative stress in the body, and then there is a healthy counterresponse that reduces that stress. But if you overdo it, that counterresponse can be overwhelmed. You'll end up with more oxidative stress, not less.

STRESSED OR DEPRESSED? EXERCISE IS RESILIENCE TRAINING FOR YOUR CELLS

"I don't have time to exercise. I'm already overcommitted and overscheduled."

"I'll exercise when I feel better. I am so stressed right now that I can't push myself to do one more hard thing."

Sound familiar? Yet it turns out that the most important time to exercise is exactly when you might not want to—when you are feeling overwhelmed. Exercise can improve your mood for up to three hours after working out[20] and can reduce stress reactivity.[21] Stress can shorten telomeres, but exercise shields telomeres from some of stress's damage. Our colleague Eli Puterman, a psychologist and exercise researcher at the University of British Columbia, has studied high-stress women, including many stressed-out caregivers. The more the women exercised, the less their stress ate away at their telomeres (see figure 17). The exercise actually buffered their telomeres from the insidious and telomere-shortening effects of stress. Even if your schedule is packed, even if you feel too exhausted to do a hard workout, find a way to slip in some exercise. For example, the two of us maintain busy schedules, but while working on this book we took walks together, thinking through the chapters aloud as we went up and down the hills of San Francisco.

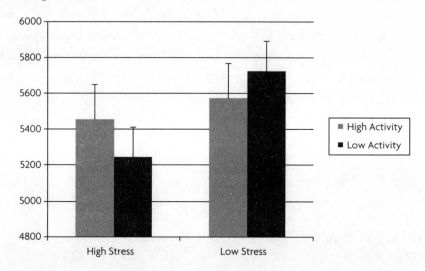

Figure 17: Physical Activity May Buffer Stress-Related Telomere Shortening. Women high on perceived stress had shorter telomeres, but only if they were relatively sedentary. If they exercised, they did not show the stress-telomere relationship.[22] The raw values (unadjusted) for telomere length in base pairs are shown here on the vertical axis.

You can probably exercise more often than you realize. But on the days when you just can't make it happen, take heart. In psychology, resilience is a kind of Holy Grail. Resilience is what keeps you bouncing back after being knocked down and lets stress slide off your shoulders without damaging your mind and body. Eli Puterman's stress research shows that telomeres can be resilient, too. The more you can practice good health habits—effective emotion regulation, strong social connections, good sleep, and good exercise—the less that stress hurts your telomeres. This is especially true if you have depression.[23] Exercise is a potent way to make your telomeres resilient, but when you can't exercise, step up other resiliency behaviors. Everything you do will help, and that's an encouraging piece of news.

TELOMERE TIPS

- People who exercise have longer telomeres than those who don't. This is true even for twins. It's the increased aerobic fitness that is most tied to good cell health.
- Exercise charges up the cell clean-up crew, so that cells have less junk buildup, more efficient mitochondria, and fewer free radicals.
- Endurance athletes, who have the best fitness and metabolic health, have long telomeres. But those telomeres are not much longer than those with moderate exercise. We don't need to aspire to extremes.
- Athletes who overexercise and burn out develop many physical issues, including risk of shorter telomeres in their muscle cells.
- If you have a high-stress life, exercise is not just good for you. It is essential. It protects you from stress-shortened telomeres.

RENEWAL LAB

IF YOU LIKE A STEADY, CARDIOVASCULAR WORKOUT...

Here is the cardiovascular workout tested in the German study, the one that showed a significant increase of telomerase.[24] It's pretty straightforward: Simply walk or run at about 60 percent of your maximum ability. You should be breathing somewhat hard, but you should still be able to maintain a conversation. Do this for forty minutes, at least three times a week.

IF YOU PREFER HIGH-INTENSITY INTERVAL TRAINING (HIT)...

This interval workout has been associated with the same gains in telomerase as the cardiovascular workout above. Plan to do it three times a week:

Cardiovascular Workout (Running)	
Warm up (easy)	10 minutes
Interval (repeat 4 times):	
Run (fast)	3 minutes
Run (easy)	3 minutes
Cool down (easy)	10 minutes

IF YOU WANT AN INTERVAL WORKOUT THAT'S LESS INTENSE...

Runners shouldn't have a monopoly on interval training. This plan is less intense but still incorporates some doable intervals. If you are out of shape, add a ten-minute warmup and cool down:

Walking Workout	
Interval (repeat 4 times):	
Walk fast (on a exertion scale of 1 to 10, be at a 6 or 7)	3 minutes
Stroll gently	3 minutes

This walking plan hasn't been tested specifically for its effects on telomeres or telomerase in any studies yet, but it certainly falls into the category of healthy exercise. One study tested this plan and found that it had a much more beneficial effect on multiple measures of fitness than just moderate, steady walking. More important, over two-thirds of the adults in this study, who were midlife or older, stayed with this walking regime for years afterward.[25]

SMALL STEPS COUNT

In addition to planned exercise, it's important to keep moving throughout the day. Activity that is woven into your daily life lifts you out of the dreaded "sedentary" category, which is linked to shorter telomeres and causes metabolic changes that can lead to more insulin resistance and inflammation.[26] So add little walks all day: Park farther away from your destination, take the stairs, or have a walking meeting. Some apps (and the iWatch) have programs that remind you to stand up every hour. Or a simple pedometer can be our daily reminder that our steps can add up.

CHAPTER EIGHT

Tired Telomeres: From Exhaustion to Restoration

Poor quality sleep, sleep debt, and sleep disorders are all linked to shorter telomeres. Of course, most of us already know that we need more sleep—the problem is figuring out how to get it. Here, we'll draw on the newest research that goes beyond the standard sleep hygiene advice and shows how both cognitive changes and mindfulness can help you get more restorative sleep. Even when you can't get more sleep, these techniques help you suffer less from the effects of sleeplessness.

Maria's sleep problems began more than fifteen years ago. She and her husband had been fighting a lot, and she'd wake up in the dark of the night, unable to stop replaying their arguments in her mind. When she consulted with a marriage and family therapist, that first bout of insomnia departed. Unfortunately, it left a door ajar, and several times a year, Maria's sleep troubles would walk back in. When they did, she'd feel too alert and aroused at night to fall asleep. She'd drift off and then wake again, often worrying about financial problems and how her sleeplessness would affect her work the next day. During the day, she felt depleted and exhausted, but her mind was racing too fast for sleep. When she attended a sleep program for insomnia, Maria was asked to track how much sleep she was actually getting. Her average minutes of sleep per night: 124.

Are *you* getting enough sleep? A quick gauge, one used by sleep researchers, is to ask yourself whether you're sleepy during the day. If you are, you need more sleep, even if your sleep loss isn't nearly as dramatic as Maria's. A better test is to ask yourself whether you fall asleep unintentionally while watching television or a movie, or while you're a passenger in a car. Many people just don't sleep enough, whether it's because of diagnosable sleep disorders, common lifestyle-related sleep problems, or being too busy; according to the National Sleep Foundation's 2014 Sleep Health Index, 45 percent of Americans say that poor or insufficient sleep affected their daily activities at least once in the last week.[1]

Telomeres need their sleep. We now know that getting enough sleep is important for healthy telomeres in all adults. Chronic insomnia is associated with shorter telomeres, particularly for people over

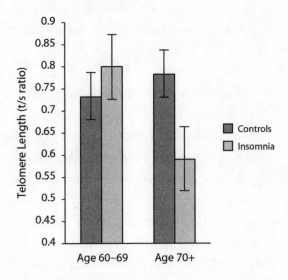

Figure 18: Telomeres and Insomnia. In men and women ages 60 to 88 years old, insomnia was related to shorter telomere length, but only in those 70 years and older. This graph shows the average telomere length from peripheral blood mononuclear cells.

seventy years old (see Figure 18).[2] In this chapter, we'll show you how good sleep protects your telomeres, buffers some of the effects of aging, regulates your appetite, and soothes the pain of some of your most stressful memories. For the newest techniques to help you get more sleep—and to help you feel better when sleep just isn't possible—read on.

THE RESTORATIVE POWER OF SLEEP

We don't usually think of sleep as an activity, but that's exactly what it is. In fact, it's the most restorative activity you can perform. You need that rejuvenating time to set your internal biological clock, regulate your appetite, consolidate and heal your memories, and refresh your mood.

Set Your Biological Clock

Do you struggle to wake up and feel alert in the morning?

Are you wide awake at bedtime?

Do you feel hungry at strange hours?

If you've answered yes to any of these questions, or your body's timing simply feels "off," you may suffer from at least slight dysregulation in a brain structure known as the suprachiasmatic nucleus, or SCN.[3] A structure of a mere fifty thousand cells, the SCN snuggles like a tiny egg within the larger nest of the brain's hypothalamus. Don't be fooled by its size, because the SCN is incredibly important. It's your body's central internal clock. It tells you when to feel tired, when to feel alert, and when to get hungry. It also drives the nightly task of cellular housekeeping, when damaged parts are swept away and DNA is repaired.[4] When your SCN is working well, you'll have more energy when you need it, deeper rest at night, and cells that are functioning more efficiently.

Like a delicate, handmade timepiece, the SCN is highly sensitive. It needs information from you to keep itself well tuned. Light signals, which are transmitted directly to the SCN through the optic nerve, allow the SCN to set itself to a proper day/night cycle. By getting light exposure during the day, and by dimming the lights at night, you keep your SCN on schedule. If you keep to regular eating and sleeping times, you also give your SCN the information it needs to inhibit the sleep drive during the day and unleash that drive throughout the night.

Control Your Appetite

Your body also depends on deep, restorative REM sleep to regulate your appetite. (REM sleep is characterized by rapid eye movements, higher heart rate, faster breathing, and more dreaming.) During REM sleep, cortisol is suppressed, and your metabolic rate increases. When you don't sleep well, you get less REM in the second half of the night, and that results in higher levels of cortisol and insulin, which stimulate appetite and lead to greater insulin resistance. In plain terms, this means that *a bad night of sleep can throw you into a temporary prediabetic state.* Studies have shown that even one night of partial sleep, or one night without enough REM sleep, can lead to elevated cortisol the next afternoon or evening, along with changes in the hormones and peptides that regulate appetite and lead to greater feelings of hunger.[5]

Good Memories, Bad Memories, and Emotions

"We sleep to remember, and we sleep to forget," says Matt Walker, a sleep researcher at the University of California, Berkeley. When you are well slept, you are better at learning and remembering. Tired people just aren't as successful at focusing their attention, so they don't take in new information as well. And sleep itself creates new connections between brain cells, which means that you're both learning and stabilizing your memory of what you've learned.

Sometimes, though, memories are painful. Sleep works its healing powers on these memories, reducing their emotional charge.

Walker has found that most of this work is accomplished during REM sleep, which shuts off some of the stimulating chemicals in your brain and allows you to split off your emotions from the content of the memory. With time, this action allows you to remember a painful experience but without an intense jolt to your mind and body.[6]

And of course, we need sleep to refresh emotionally. If you don't already know that sleep loss makes you more irritable, ask your family or colleagues. They'll quickly confirm this for you. When you are not well slept, you have a physiological and emotional stress response that's measurably bigger.[7] You can even get giddy or giggly more easily.[8] Sleep deprivation makes *all* emotions more intense. This is, perhaps, a reason Maria felt so hyperaroused and jumpy.

HOW MANY HOURS OF SLEEP DO TELOMERES NEED?

As scientists have realized that sleep is crucial to your mind, your metabolism, and your mood, they have increasingly included measurements of telomeres in their sleep studies. Researchers have looked at how sleep length affects telomeres in different populations, and the same answer keeps coming up: Long sleep means long telomeres.

Getting at least seven hours of sleep or more is associated with longer telomeres, especially if you are older.[9] The famous Whitehall study of British civil servants found that men who slept five hours or fewer most nights had telomeres that were shorter than men who slept more than seven hours.[10] This finding was after adjusting for other factors such as socioeconomic status, obesity, and depression. Seven hours of sleep appears to be the cutoff point for telomere health. Get less than seven, and telomeres start to suffer. If you're one of those rare people who need very little sleep (about 5 percent of the population needs only five or six hours of sleep per night), this cutoff point doesn't apply to you. Then again, if you feel terrible without eight or nine hours of sleep, don't try to scrape by with seven. Get those extra hours. And remember that rule of thumb, which offers highly customized sleep advice: *If you feel sleepy during the day, you need more sleep at night.*

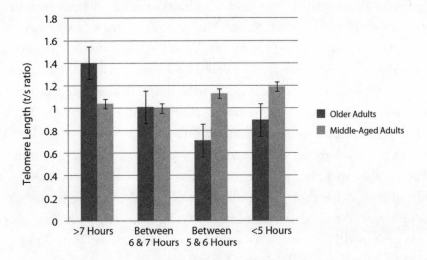

Figure 19: Telomeres and Hours of Sleep. Older adults who only get five or six hours of sleep a night have shorter telomeres. If they get more than seven hours of sleep, their telomere length is similar to younger adults.[11]

It's Not Just Hours in Bed: Sleep Quality, Regularity, and Rhythm

As you keep the seven-hour goal in mind, try not to become obsessed with it, because it's not just hours that matter. Think back to the past week and how well you've slept. How would you rate your sleep quality for the last seven days? Was it very good, fairly good, fairly bad, or very bad? The answers to this straightforward question have been scientifically linked to your telomere health. The closer you are toward the "very good" end of the scale, the healthier your telomeres are likely to be. In several studies examining sleep quality, people who gave themselves high ratings for sleep quality had longer telomeres.

Good sleep appears to be especially protective as we get older,

193

buffering the natural age-related decline in telomere length. In one study age was not related to shorter telomere length in those with great sleep quality.[12] When sleep quality remains good, telomeres stay pretty stable across the decades.

Good sleep quality also protects the telomeres in your immune system's CD8 cells. When these cells are young, they attack viruses, bacteria, and other foreign invaders. Your body is constantly fighting off threats, but when you're protected by a vigorous army of immune cells, including CD8 cells, you hardly notice those threats. That's because the invaders are being surrounded and destroyed. These CD8 cells are part of an incredibly effective defense system. Until, that is, their telomeres shorten and they start to grow old. Then they are less able to fight off foreign bodies in your bloodstream; that's why people with shorter telomeres in CD8 cells are more likely to catch a cold virus. Short telomeres in CD8 cells can, with time, lead to systemic inflammation, as we've mentioned earlier. Aric Prather, PhD, a sleep researcher at the University of California, San Francisco, has found that women who rated their quality of sleep as poor were more likely to have short telomeres in their CD8 cells; excessive daytime sleepiness was also a predictor of shorter telomere length. Women with high stress were the most vulnerable to the effects of poor-quality sleep.[13]

Sleep length and sleep quality are important. Now add sleep rhythm to the list. Keeping a good sleep-wake rhythm—going to bed and waking up at regular times—may be critical to your cells' ability to regulate telomerase. In one study, scientists removed the "clock genes" from mice. Although normal mice show higher telomerase in the morning and lower telomerase at night, the mice without the clock gene did not show this telomerase diurnal rhythm, and their telomeres shortened. Then the same investigators turned to humans whose work schedules had, effectively, broken their internal clocks. Emergency room physicians who pulled night-shift duty also lacked this normal rhythm of telomerase.[14] This study was small, but it suggests that good sleep-wake rhythms may be critical in helping to maintain the rhythm of telomerase activity best for keeping your telomeres replenished.

HELP FOR SLEEP PROBLEMS: COGNITION AND METACOGNITION

Some of us need to be convinced that sleep is vital for health, but Maria did not. Driven by desperation, she attended a clinic that was experimenting with a novel approach to sleep problems.

Insomnia is characterized by some universal experiences: feeling too alert to fall sleep; trying too hard to sleep; and, especially, the common habit of rehearsing the past or worrying about the future. In order to sleep, we need to feel physically and psychologically safe. But at night, small worries can morph into large, looming threats, making it hard to feel secure enough to sleep. Usually these threats are, as Elissa's father used to say, "mere demons of the night" that disappear in daylight. He was right. Nighttime can turn manageable worries, problems that can be solved by day, into a chain of catastrophes that are replayed in a stupor of tired rumination.

But there is a second layer of worries that can arise. This tricky layer is made up of worries about insomnia and its effects. They include:

- "I can't function tomorrow without a good's night sleep."
- "I should be able to sleep as soundly as my bed partner does."
- "I'm going to look terrible tomorrow."
- "I'm going to have a nervous breakdown."

These thoughts can push an episode of tossing and turning into full-blown insomnia, and they can color the negative emotions you might feel the next day with an even darker tint.

One method that's been shown to help with this second layer of thoughts is to examine them directly. Like the demons of the night, your thoughts about sleep are usually much less foreboding and dramatic when you examine them in the light of day. They're what we call "cognitive distortions," and they mostly aren't true. Challenge these thoughts, and you'll find more accurate statements emerging:

- "Although I don't function as well without sleep, I can still manage to get my tasks done."
- "My partner's sleep needs are not the same as mine."
- "I look pretty good" or "Thank goodness for makeup!"
- "I'm going to be okay."

Dr. Jason Ong directed the sleep program that Maria attended. Cognitive behavioral therapy is the best known treatment for insomnia so far, as it challenges your thoughts about insomnia. At the same time, Jason also noticed that when sleep therapists challenged their patients' thoughts, some patients felt a bit bullied, as if the doctor was telling them what to think. Or they felt as if they were on opposite sides of a debate, with opposing arguments flying back and forth.

So in Dr. Ong's workshops, patients practice the usual good sleep behaviors that most doctors prescribe—getting out of bed when they are not able to sleep, waking up at the same time every morning, not trying to make up for lost sleep with naps—but instead of telling patients to think differently, the therapists encourage them to watch their thoughts from a distance. This is, again, a form of mindfulness. At the clinic, patients like Maria learn different forms of meditation, including moving meditations (for example, walking slowly while paying attention to each step) and more traditional meditations (sitting quietly and noticing the breath). They are encouraged to accept their thoughts about insomnia, and then to let those thoughts go. Meditation isn't used as a way to make people feel sleepy—it's a method for promoting awareness of that second layer of thoughts that make insomnia so much worse. It defuses those thoughts.

196

It can take a while to change your relationship to your thoughts. Maria stuck with the meditation program for six weeks without seeing much improvement. Finally she expressed her frustration. She said, "During my meditations, I have been trying to clear my mind and sometimes I can keep it blank for a while, but [the thoughts] always come back."

Dr. Ong suggested that Maria stop trying to exercise so much power over her mind. He asked her to consider what would happen if she simply let her thoughts run their course. "It is not the thoughts you are trying to control, but instead it is letting go of the effort of forcing these thoughts to go in a certain direction," he explained.

Maria mulled this over, and tried the meditation again with this new, less forceful approach. The following week, her anxiety levels downshifted. She felt less worked up before going to bed at night, and at the next workshop she was noticeably more relaxed. "For a long time I thought I had to get rid of my thoughts to sleep better. It's funny that once I stopped trying to make that happen, my sleep seemed to get better," she reported. Over the next few weeks, she nearly doubled her average sleep time—not a total cure, but a significant improvement. Her doctors predicted that as she continued to practice mindfulness, she would make further gains.[15]

Ong tested out his eight-week mindfulness-based treatment for insomnia. The program, officially known as MBTI, was compared to a group who simply wrote down their sleep times and levels of arousal. Those in the MBTI group had greater decreases in insomnia, and within six months, 80 percent showed improvement in their sleep.[16]

FRESH STRATEGIES FOR MORE SLEEP

What about everyone else, including those of us who don't have chronic insomnia but who could use help getting a little more sleep? Following are some suggestions.

Give Yourself the Gift of Protected Transition Time

Your mind is not a car engine. You can't run it high speed right up until bedtime, doing work, exercise, chores, or tending to children, and then expect to switch it off and drop into slumber. It doesn't work that way. Biologically, *your brain is more like an airplane.* You need a slow descent into sleep, landing as gently as possible. So give yourself the gift of transition time between work and sleep, a sleep routine or ritual that lets you wind down. The smoother the transition, the less jolted you'll feel when you land.

Even five minutes of transition time can make a difference. Begin by unplugging. Turn off your phone or set it to airplane mode; let your body have a break from instant responding. If you have the willpower, leave your phone in a different room entirely. By setting aside phones and other screens, you minimize the number of stressors that can feature on your mind's IMAX screen of nighttime worries. You already have enough stress to contend with, given the natural human tendency to ruminate and rehearse worries at night. (In the next section, you'll see that screens are also sources of blue light, which keeps you awake.) After you've turned off the screens, perform a quiet, pleasant activity—not to make yourself sleepy but to create a transition period of calm and comfort. Some people like to read or knit or even open up a stress-relieving coloring book designed for adults. (You can find an adult coloring page in this chapter's Renewal Lab.) You can listen to an audio meditation or music that relaxes you.

Blue Light Suppresses Melatonin

There was a worldwide sleep debt even before our current addiction to screens. But now there are extra challenges to sleep. Do you bring smartphones, tablets, or other screens into your bedroom? The blue light from screens can suppress melatonin, the sleepiness hormone. In a study by sleep researcher Charles Czeisler and colleagues, people

198

who used an e-reader immediately before bed released around 50 percent less melatonin than people who read from a print book.[17] The people using an e-reader took longer to fall asleep, had less REM sleep, and felt less alert in the morning.

Try to avoid screens for an hour before bedtime. If you can't, try using smaller screens, and holding them farther from your eyes, to help minimize blue light exposure. Liz uses a free software program called f.lux that matches a screen's light to the time of day, so that the blue light fades into yellow as you head into evening. You can download it at https://justgetflux.com. Apple computer's new operating system 9.3 has Night Shift, a program that automatically shifts from blue to yellow at night.

All light suppresses melatonin, though, so keep things as dark as you can. Look around your room at night. Where do you see light? Minimize light from windows and digital clocks. Wear an eye mask, and let the melatonin flow.

NOISE, HEART RATE, AND SLEEP

We all come with different settings for sleep. Some just aren't bothered by noise and some are. People with a particular pattern of brain activity, whose EEGs show the bursts of brain waves known as spindles, appear to be more resilient to nighttime noises.[18] For the rest of us, hearing sounds like car horns or sirens speeds up the heart rate and disrupts the sleep cycle.[19] If you are highly sensitive to your surroundings, you need to control your exposures. The more completely you can tune out of your environment, the more you can feel safe from the intrusion of noise, and the deeper you'll sleep. Earplugs are a good place to begin.

Synchronize Your Brain with Your Internal Clock

Your suprachiasmatic nucleus, the brain's clock, is trying to keep circadian rhythms on track. Help it along by eating and sleeping at regular times. This regularity will help your brain know when to release melatonin, and it will help your cells know when it's time to repair DNA and perform other restorative functions. Regular mealtimes and sufficient sleep also leads to greater insulin sensitivity, which helps you burn fat more efficiently.

Sleep Debt Is Not a Blame Game

People lose sleep at predictable times: after a baby is born, when their partner goes through a snoring phase, when they feel depressed or stressed, when hot flashes hit, or when first adapting to age-related changes in sleep. These events are usually temporary. They happen, and they pass. But today's epidemic levels of sleep loss are not caused by these events. Most sleep loss is caused by "voluntary sleep curtailment," otherwise known as sleep procrastination, otherwise known as not getting yourself into bed early enough.

You might be responding as I (Elissa) did when I heard that term: "I'm not volunteering to lose sleep—it's just that I have too much to do." But instead of mentally preparing your defense, remind yourself that sleep loss can't be resolved by playing a blame game. Just gently remind yourself that unless you're a new parent or a caregiver, your bedtime is one of the few areas of sleep you *can* control. Take advantage of this power and go to sleep earlier. (An exception: Severe insomnia and age-related sleep changes don't respond to an earlier bedtime. In these cases, going to bed earlier can boomerang on you and make it harder to get quality sleep all the way through the night.)

Treat Sleep Apnea and Snoring

Severe sleep apnea, the repeated cessation of breathing during sleep, has been linked to shorter telomeres in adults.[20] The cellular effects of sleep apnea may even be transmitted in the womb. In a sample of pregnant women, 30 percent responded to a sleep assessment with answers suggesting symptoms of apnea. When these women's babies were born, the telomeres in the babies' cord blood were shorter.[21] The same was true for women who snored. And here's some bad news for the many snorers out there: More time spent snoring is also linked to telomere shortness, at least in a large sample of Korean adults.[22] If you suspect you have sleep apnea, get tested and then take advantage of the effective new treatments, which are more comfortable than traditional CPAP machines, the ones that apply air pressure through a mask.

SLEEP IS A GROUP PROJECT

You probably know a few people who get enough sleep. You can spot them easily: They're the ones with bright eyes and skin, the ones who don't constantly complain about how tired they are, who don't always have a Starbucks Grande in one hand, the ones who don't wonder why they feel hungry at strange times of the day. What do these people have that the rest of us don't? Well, a few things. They may have a partner who encourages good sleep—and who suggests leaving the phone in the kitchen to recharge overnight. They may have colleagues who don't drop emergency e-mails at 10:00 p.m. They may have children who go to bed and stay there!

What we're saying here is that sometimes sleep is a group project. We have to support one another in reducing sleep procrastination, in going to bed earlier, in not doing business late at night. As the saying goes, be the change you wish to see in the world. Make a pact with your spouse to leave a few minutes at night to transition out of the stress mind-set. Make another pact with your colleagues not to send late-night messages (if you have to write them at night, save them in your drafts folder until morning). You can't tell your children not to have the kinds of nightmares that send them running to you at 2:00 a.m., but you can set an example of what good sleep habits look like in adulthood.

TELOMERE TIPS

- With sufficient sleep you will feel less hungry, less emotional, and lose fewer telomere base pairs.

- Telomeres like at least seven hours of sleep. Many strategies can help us boost our sleep quality, some as simple (but hard) as removing electronic screens from our bedroom.

- Try to minimize the effects of sleep apnea, snoring, and insomnia. These problems are common later in life. And when insomnia visits, use comforting thoughts to soften the alarming ones. If you have severe insomnia, cognitive-behavioral therapy can help.

RENEWAL LAB

FIVE BEDTIME RITUALS

Bringing restfulness into your bedtime space promotes a better night's sleep. Begin by making the next day's to-do list. Then put the list aside. That way, you'll feel more peaceful about tomorrow, and you'll leave behind some of the mental effort that keeps you in vigilance-anticipation mode. After that, you're ready for a bedtime ritual. Here are five rituals to support maximum tranquility and relaxation:

1. Spend five minutes in transition: breathing, meditating, or reading. The centuries-old practice of reading a book before bed can also help transition from an overly active mind to a state of absorbed attention. The transfer of focus away from self and to the content of the book can quiet the mind—provided that the book is not too exciting, of course.

2. Listen to soothing music. Soothing music calms your nervous system and your mind, and it sends a signal to start transitioning into a state of rest. The Spotify app offers several bedtime playlists, including "Bedtime Bach" (for lovers of classical music), "Best Relaxing Spa Music" (if you prefer New Age), and many soporific options under "sleep" such as "Sleep: Into the Ocean" (if you like nature sounds).

3. Set a mood for relaxation. Use essential oils, light a candle, and dim the lights. When our environments are restful and peaceful, so are we. Calming scents like lavender, cedar, or sandalwood are soothing for your entire system and brain. Reducing artificial light, and then turning off the lights completely, is a must for becoming restful enough to drift to sleep.

4. Brew warm herbal tea an hour or more before bed. A warm, scented mug of tea will help you wind down from your day. Try making your own blend of herbal tea from chamomile, lavender, rose petals, and a slice of fresh lemon or ginger. Don't drink the tea right before bed, or your sleep might be disrupted by a bathroom break.

5. Perform bedtime stretches or do some gentle yoga. Simple head and neck rolls will help melt the tension and anxiety from the day. For a more structured bedtime yoga routine, try this one. You can do it on a yoga mat or right on your bed:

Gentle Rolling of the Head and Neck: Start by slowly and gently rolling your head and neck in a clockwise motion while inhaling and exhaling long, deep breaths. Bring your attention particularly to the exhale, as this helps you let go of any stress incurred during your day. After one minute, gently switch directions. Roll your head and neck counterclockwise for one minute.

Forward Bend: Sit with a straight spine and your legs fully stretched out before you and parallel with your mat or bed. Pause here and take a long, deep inhale. On the exhale, start to bend or hinge at your waist, stretching your hands toward your feet. You can rest your hands on your calves, on the bed or mat beside your thighs, or on the tops of your feet. Allow yourself to be in this modified forward bend for at least three breaths or longer. When you're ready, slowly and mindfully activate your core to roll your spine into an upright position, long and straight, right where you began.

Child's Pose: The perfect send-off to bed is simply lying and breathing in child's pose (see Figure 20). Child's pose is a traditional resting posture in yoga and allows your entire system and body to relax. Start in a seated kneeling position. Take one long inhale, and on the exhale hinge or bend forward, bringing your head down to the mat or bed. Rest feeling fully supported in child's pose for several minutes, mindfully following your breath. When you're ready, return to the original kneeling position.

You are now ready for a good night's sleep.

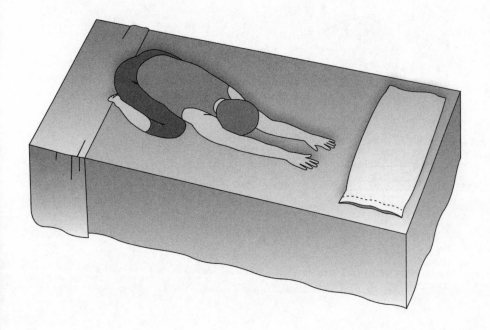

CHAPTER NINE

Telomeres Weigh In: A Healthy Metabolism

Your telomeres care how much you weigh—but not as much as you might imagine. What really appears to matter to telomeres is your metabolic health. Insulin resistance and belly fat are your real enemies, not the pounds on the scale. Dieting affects telomeres, both for good and for ill.

My (Elissa's) friend Peter is a genetic researcher and athlete who competes in Olympic-distance triathlons. He is muscular and burly, and his handsome face glows from his daily exercise. Peter has a huge appetite, but he works hard to keep himself from eating too much. I have spent a lot of time studying the psychology of eating, so I asked him what it's like to think so much about *not* eating:

I would have been an awesome hunter-gatherer. I can sniff out food in a second, especially sweets. At work, it's a joke: when food appears, so does Peter. I know where people will put food out—one person has a candy jar she fills periodically, another puts a plate of food out on a counter near her office, and lots of people put snacks or leftovers from parties or their kids' Halloween stash on the table in the kitchen.

I try to avoid seeing the food. When I meet with the woman who has the candy bowl, I try hard not to look at it (she's my boss, and I should be listening to her, but sometimes I'm thinking about not looking at the candy). When I get up to go to the bathroom, I choose a route that won't take me near the kitchen. But that means I can't even pee without thinking about food: Will I go by the kitchen to see if something is there? Or will I be strong and take a different route? I have to answer that question pretty much every time I leave my desk, because it's so easy to choose a route that will take me by a place where there might be food.

My plans to eat well don't always work. For instance, I often bring a healthy salad to work, but I don't always eat it, because I have to store it in the kitchen. I'll be on my way to get a salad, and get intercepted by the pound cake that someone put on the kitchen table. I end up eating a pound of cake—isn't that why they call it a pound cake?—while the salad sits, wilting and forgotten.

As Peter has discovered, it's hard work to think about food all the time, and even harder to lose weight. However, there is hopeful news for Peter and everyone else who struggles with weight, diet, and stress: It is not necessary, or even healthy, to think so much about food and caloric intake. That's because your telomeres care about your weight, but not as much as you might think.

IT'S THE BELLY, NOT THE BMI

Does eating too much shorten your telomeres? The quick and easy answer is *yes*. The effect of excess weight on telomeres is real—but it's not nearly as striking as the relationship between, say, depression and telomeres (which is around three times larger).[1] The weight effect is small and probably not directly causal. This finding may come as a surprise to people like Peter, who devote a huge chunk of their mental resources to the effort of eating less. It may be a bit shocking to *everyone* who's heard the message that weight loss is the most urgent goal in public health. Yet being overweight (and not obese) is, surprisingly, not linked strongly to shorter telomeres (nor is it strongly linked to mortality). Here's the reason: weight is a crude stand-in measure for what really matters, which is your metabolic health.[2] Most obesity research relies on the measure of body mass index (BMI, a measure of weight by height), but this does not tell us much about what really matters—how much muscle versus body fat we have, and where the fat is stored. Fat stored in the limbs (subcutaneously, so under the skin but not in the muscle) is different and maybe even protective, while fat stored deep inside, in the belly, liver, or muscles, is the real underlying threat. We are going to show you what it means to have poor metabolic health and show you why dieting may not be the way to get healthier.

Growing up, Sarah impressed her friends and family with her appetite. "I'd eat an Italian sub sandwich as an after-school snack, washed down by two glasses of sweet iced tea, and I'd never gain weight," she recalls wistfully. Sarah ate her way through high school and college; throughout a charmed early adulthood, she was slim. Until, suddenly, she wasn't. She was eating the same things and exercising the same amount (which was very little). Her upper body and legs were still trim, but her pants stopped fitting. Sarah had developed a belly. "I look like a strand of spaghetti with a meatball in the middle," she says now. She's worried, because both her parents take medication for high levels of bad cholesterol. After three decades of feeling effortlessly healthy, Sarah is wondering if she is going to join her parents in line at the pharmacy.

She's right to be worried, and it's not just her cholesterol levels that are at stake. Sarah's body type, where the weight is overrepresented at the belly, is closely associated with poor metabolic health. This is true *no matter how much you weigh*. It's true for people who carry a huge beer belly, and it's true for Sarah, whose BMI is normal but whose waist circumference is bigger than her hips.

209

When we say a person has poor metabolic health, we generally mean that he or she has a package of risk factors: belly fat, abnormal cholesterol levels, high blood pressure, and insulin resistance. Have three or more of these risk factors and you get labeled with "metabolic syndrome," a precursor to heart problems, cancer, and one of the greatest health threats of the twenty-first century: diabetes.

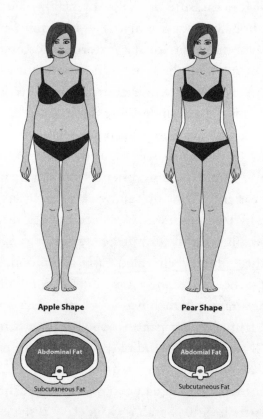

Apple Shape   **Pear Shape**

Figure 22: Telomeres and Belly Fat. Here you see what it means to have excessive fat around the waist, an apple shape (reflecting high intra-abdominal fat, measured by a greater waist-to-hip ratio, or WHR), versus more fat in the hips and thighs, a pear shape (smaller WHR). Subcutaneous fat, found under the skin and in limbs, carries fewer health risks. High intra-abdominal fat is metabolically troublesome and indicates some level of poor glucose control or insulin resistance. In one study, greater WHR predicts 40 percent greater risk for telomere shortening over the next five years.[3]

BELLY FAT, INSULIN RESISTANCE, AND DIABETES

Diabetes is a global public health emergency. The list of its long-term effects is long and chilling: heart disease, stroke, vision loss, and vascular problems that can require amputations. Worldwide, more than 387 million people—that's nearly 9 percent of the global population—have diabetes. That includes 7.3 million in Germany, 2.4 million in the United Kingdom, 9 million in Mexico, and a colossal 25.8 million people in the United States.[4]

Here's how type 2 diabetes develops: in a healthy person, the digestive system breaks food down into glucose. The beta cells in the pancreas make a hormone, insulin, which is released into the bloodstream and allows glucose to enter the body's cells to be used as fuel. In a wonderfully tidy system, insulin binds to receptors on the cells, like a key fitting into a lock. The lock turns, the door opens, and glucose can enter the body's cells. But too much belly or liver fat can cause your body to become insulin resistant, meaning that cells don't respond to insulin the way they should. Their "locks"—the insulin receptors—gum up and stick; the key no longer fits as well. It's harder for glucose to enter the cells. The glucose that can't get in through the door remains in the bloodstream. Glucose builds up in the blood even as your pancreas churns out more and more insulin. Type 1 diabetes is related to failure of the beta cells in the pancreas; they can't produce enough insulin. You're at risk for metabolic syndrome. And if your body can't keep glucose in the normal range, diabetes results.

HOW SHORT TELOMERES AND INFLAMMATION CONTRIBUTE TO DIABETES

Why do people with belly fat have more insulin resistance and diabetes? Poor nutrition, inactivity, and stress are all associated with belly fat and higher levels of blood sugar. But people with belly fat develop shorter telomeres over the years,[5] and it's very possible that these short telomeres worsen the insulin resistance problem. In a Danish study of 338 twins, short telomere length predicted increases in insulin resistance over twelve years. Within twin pairs, the twin with shorter telomeres developed higher insulin resistance.[6]

There's also a well-established connection between short telomeres and diabetes. People afflicted with inherited short telomere syndromes are much more likely to develop diabetes than the rest of the population. Their diabetes comes on early and strong. Other evidence comes from Native Americans, who are at a high risk of diabetes for a variety of reasons. When an Native American has short telomeres, he or she is twice as likely to develop diabetes over the course of five years than other members of this ethnic group with longer telomeres.[7] A meta-analysis across studies of around seven thousand people shows that short blood cell telomeres predict future onset of diabetes.[8]

We even have a glimpse into the mechanism that causes diabetes and can see what's happening in the pancreas. Mary Armanios and her colleagues have shown that when a mouse's telomeres are shortened throughout its body (through a genetic mutation), its pancreas's beta cells are not able to secrete insulin.[9] And the stem cells in the pancreas become exhausted; they run out of telomere length and can't replenish the damaged pancreatic beta cells that should have been doing the work of insulin production and regulation. These cells die off. Type 1 diabetes steps in and begins its malevolent work. In the more common type 2 diabetes, there is some beta cell dysfunction, and so short telomeres in the pancreas may play some role there as well.

In an otherwise healthy person, the pathway from belly fat to diabetes may also be traveled via our old enemy, chronic inflammation. Abdominal fat is more inflammatory than, say, thigh fat. The fat cells secrete proinflammatory substances that damage the cells of the immune system, making them senescent and shortening their telomeres. (Of course, one hallmark of senescent cells is that they can't stop sending out proinflammatory signals of their own. It's a vicious cycle.)

If you have excess belly fat (and more than half the adults in the United States do), you may be wondering how you can protect yourself—from inflammation, from short telomeres, and from metabolic syndrome. Before you go on a diet to reduce belly fat, read the rest of this chapter; you may decide that a diet will only make things worse. And that's fine, because soon we will suggest some alternate ways to improve your metabolic health.

DIETING IS DISAPPOINTING (WHAT A RELIEF)

There is a relationship between dieting, telomeres, and your metabolic health. But as in all things related to weight, it's complicated. Here are some results of research into weight loss and telomeres:

- Weight loss leads to a slowdown in the normal attrition rate of telomeres.
- Weight loss has no effect on telomeres.
- Weight loss encourages telomeres to lengthen.
- Weight loss leads to shorter telomeres.

It's a mind-bending set of findings. (In that final study, people who underwent bariatric surgery had shorter telomeres one year after their procedure, though this effect was possibly from the physical stress of the surgery.)[10]

We think that these mixed results are telling us that, once again, it's not really the weight that matters. Weight loss is only a crude stand-in for positive changes to underlying metabolic health. One of those changes is the loss of belly fat. Lose weight overall, and you'll inevitably take a bite out of that "apple," and this may be more true if you are increasing your exercise rather than just reducing calories.

Another positive change is improved insulin resistance. One study followed volunteers for ten to twelve years; as the people in the study gained weight (as people tend to do), their telomeres got shorter. But then the researchers examined which mattered more, weight gain or the insulin resistance that often comes with it. It was insulin resistance that carried the weight, so to speak.[11]

This idea—that improving your metabolic health is more important than losing weight—is vital, and that's because repeated dieting takes a toll on your body. There are some internal "push back" mechanisms that makes it hard for us to keep weight off. Our body has a set point that it defends, and when we lose weight, we also slow our metabolism in an effort to regain the weight ("metabolic adaptation"). While this is well known, we didn't know how dramatic this adaptation could be. There is a tragic lesson here from the brave volunteers who have joined the reality TV show *The Biggest Loser*. For this show, very heavy people compete to lose the most pounds over 7.5 months, using exercise and diet. Dr. Kevin Hall and his colleagues from the National Institutes of Health decided to examine how this rapid massive weight loss affected their metabolism. At the end of the show, they had lost 40 percent of their weight (around 58 kg). Hall checked their weight and metabolism again six years later. Most had regained weight, but they kept an average of 12 percent weight loss. Here is the hard part: at the end of the program, their metabolism had slowed so that they were burning 610 fewer calories per day. By six years, despite weight regain, their metabolic adaptations had become even more severe, where they were burning around 700 calories fewer than their baseline.[12] Ouch. While this is an example of extreme weight loss, this metabolic rate slowing happens to a lesser extent whenever we lose weight, and, apparently, even when we regain it.

In the phenomenon known as weight cycling (or "yo-yo dieting"), dieters gain pounds and lose them, and gain and lose, and so on. Fewer than 5 percent of people who are trying to lose weight can stick to a diet and maintain the weight loss for five years.

The remaining 95 percent either give up or become weight cyclers. Weight cycling has become a way of life for many of us, especially women; it's what we talk about; it's how we laugh together. (An example: "Inside me there's a skinny woman crying to get out, but I can usually shut her up with cookies.") Yet weight cycling appears to shorten our telomeres.[13]

Weight cycling is so unhealthy, and also so common, that we feel strongly that everyone should understand it. Weight cyclers restrict themselves for a while and then, when they fall off the wagon, tend to indulge in treats and other unhealthy foods. This intermittent cycling between restriction and indulgence is a real problem. What happens to animals when they get junk food all the time? They overeat and get obese. But when you withhold junk food most of the time, giving it to them only every few days, something even more disturbing happens. The rats' brain chemistry changes; the brain's reward pathways start to look like the brains of people who are suffering from drug addiction. When the rats don't get their sugary, chocolaty rat junk food, they develop withdrawal symptoms, and their brains release the stress chemical CRH (short for corticotropin-releasing hormone). The CRH makes the rats feel so bad that they are driven to seek the junk food, to get relief from their stressed state of withdrawal. When the rats finally do get the chocolaty stuff, they eat it as if they will never have the chance again. They binge.[14]

Sound like anyone you know? Or like Peter eating pound cake on his way to eat a healthy salad for lunch? Studies of obese people suggest a similar compulsive aspect of overeating, with dysregulation in the brain's reward system.

Dieting can create a semiaddictive state, and it's also just plain stressful. Monitoring calories causes cognitive load, meaning that it uses up the brain's limited attention and increases how much stress you feel.[15]

Think of Peter, spending years trying to eat fewer sweets and calories. Obesity researchers have a name for this kind of long-term dieting mentality: cognitive dietary restraint. Restrainers devote a lot of their time to wishing, wanting, and trying to eat less, but their actual caloric intake is no lower than people who are unrestrained. We asked a group of women questions such as "Do you try to eat less at mealtimes than you like to eat?" and "How often do you try not to eat between meals because you are watching your weight?" The women who answered in ways that revealed a high level of dietary restraint had shorter telomeres than carefree eaters, regardless of how much they weighed.[16] It's just not healthy to spend a lifetime thinking about eating less. It's not good for your attention (a precious limited resource), it's not good for your stress levels, and it's not good for your cell aging.

Instead of dieting by restricting calories, focus on being physically active and eating nutritious foods—and in the next chapter we will help you choose the foods that are best for your telomeres and overall health.

SUGAR: NOT A SWEET STORY

When we want to spot the parties responsible for metabolic disease, we point a finger straight at highly processed, sugary foods and sweetened drinks.[17] (We're looking at you, packaged cakes, candies, cookies, and sodas.) These are the foods and drinks most associated with compulsive eating.[18] They light up the reward system in your brain. They are almost immediately absorbed into the blood, and they trick the brain into thinking we are starving and need more food. While we used to think all nutrients had similar effects on weight and metabolism— "a calorie is a calorie" —this is wrong. Simply reducing sugars, even if you eat the same number of calories, can lead to metabolic improvements.[19] Simple carbs wreak more havoc on our metabolism and control over appetite than other types of foods.

EXTREME CALORIC RESTRICTION: IS IT GOOD FOR TELOMERES?

You're in a cafeteria, standing in line with your tray. When you get to the front of the line, you notice that everyone is using pairs of tongs to select tiny morsels of food, which they carry over to a scale and weigh carefully. Once they are satisfied with the number of grams of food they have chosen, they take their trays—which bear much less food than you'd normally choose for yourself—to a table and sit down. You join them and watch them eat their meager lunches. When their plates are empty, they say, "Still a bit hungry," and smile.

Why are these people weighing out small portions of food? Why are they smiling when they're hungry? This is a thought exercise—no such cafeteria exists in the real world—but it reflects the habits of people who believe that by restricting their calories to 25 or 30 percent less than a normal healthy intake, they will live longer. People who practice caloric restriction teach themselves to have a different reaction to hunger. When they feel the pang of an empty stomach, they don't feel stressed or unhappy. Instead, they say to themselves, *Yes! I'm reaching my goal.* They are incredibly good at planning and thinking about the future. For example, a caloric restriction practitioner in one of our studies was eagerly organizing his 130th birthday, even though he was only around sixty years old at the time.[20]

If only these people were worms. Or mice. There is little doubt that extreme caloric restriction extends the longevity of various lower species. In at least some breeds of mice on restricted diets, telomeres appear to lengthen. They also have fewer senescent cells in the liver, an organ that is one of the first places senescent cells will build up.[21] Caloric restriction can improve insulin sensitivity, too, and reduce oxidative stress. But it's harder to pinpoint the effects of caloric restriction on larger animals. In one study, monkeys who ate 30 percent fewer calories than normal had a longer healthspan and longer life—but only when compared to a control group of monkeys who ate a lot of sugar and fat. In a second study, monkeys on a similarly restricted diet were compared to monkeys who ate normal portions of healthy food. Those monkeys did not have more longevity, though they stayed in the healthspan a bit longer. Adding to the uncertainty is that in both studies, the monkeys ate in solitude. Monkeys are highly social animals; in the wild, they eat together. Having them eat in circumstances that were abnormal, and quite possibly stressful, could have affected the outcome in ways we don't yet understand.

For now, it looks as if caloric restriction has no positive effect on human telomeres. Janet Tomiyama, now a psychology professor at UCLA, conducted a study during her postdoctoral fellowship at UCSF. She managed to round up a group of people from across the United States who were successful at long-term caloric restriction for an intensive study where she also examined telomeres in different blood cell types. (As you may imagine, such people are rare.) To our surprise, their telomeres weren't any longer than the normal or even the overweight control group. In fact, their telomeres tended to be slightly shorter in their peripheral mononuclear blood cells, which are types of immune cells that include the T-cells. Another study looked at rhesus monkeys who were restricted to 30 percent fewer calories than a normal rhesus monkey diet. The researchers measured telomere length in various tissues—not just blood, which is the typical source of telomere measurements, but also in fat and muscle. Once again, there were no differences in telomere length in the calorie-restricted monkeys—not in any of the cell types.

Thank goodness. Most people can't practice extreme calorie restriction, and few people want to. As one of our friends said, "I'd rather eat good dinners until I'm eighty than starve until I'm one hundred." He's got a point. You do not have to suffer to eat in a way that is good for your telomeres and good for your healthspan. To learn more, turn to the next chapter.

TELOMERE TIPS

- Telomeres tell us not to focus on weight. Instead, use your level of belly protrusion and insulin sensitivity as an index of health. (Your doctor can measure your insulin sensitivity by testing your fasting insulin and glucose.)
- Obsessing about calories is stressful and possibly bad for your telomeres.
- Eating and drinking low-sugar, low-glycemic-index food and beverages will boost your inner metabolic health, which is what really matters (more than weight).

RENEWAL LAB

SURF YOUR SUGAR CRAVINGS

Cutting back on sugar may be the single most beneficial change you can make to your diet. The American Heart Association recommends limiting added sugar to nine teaspoons a day for men and six teaspoons for women, but the average American has almost twenty teaspoons a day. A high-sugar diet is associated with more belly fat and insulin resistance, and three studies have found a link between shorter telomeres and drinking sugared beverages. (In the next chapter we'll talk about sugary drinks in more detail.)

When you get a craving for sugar (or any other food that isn't good for you), you need a tool to help you cope. Cravings are strong, and they're backed up by dopamine activity in the reward center of the brain. Fortunately, cravings are impermanent. They will pass. Psychologist Alan Marlatt has applied the idea of "Surfing the Urge" to help people with addictions resist their cravings until those cravings dissipate. Andrea Lieberstein, a mindful eating expert, has found this practice works even better for food cravings when adding a heart focus to the end, taking the edge off of the craving with feelings of compassion and kindness.

Here's how to surf your cravings:

SURF YOUR CRAVINGS

Sit comfortably and close your eyes. Picture the snack or treat you're craving: Conjure up its texture, its color, its smell. As the image becomes vivid, let yourself feel your craving. Let your attention wander throughout your body to observe the nature of this craving.

Describe this craving to yourself. What are its sensations and its qualities? What are the shapes, sensations, and any thoughts or feelings associated with it? Where is it located in your body? Does it change as you notice it, or as you exhale? Feel any discomfort. Remind yourself this is not an itch that needs to be scratched. This is a feeling that changes and will pass. Perhaps imagine it as a wave that builds, crests, and dissipates back into the ocean. Breathe into the sensation, and let it release tension, as you notice the waves falling back gently.

You might put your attention and your hand on your heart, imagining a sense of warmth and kindness flowing outward from your heart. Let this sense of warmth spread throughout your body, enveloping the sense of craving with loving-kindness. Take a moment just breathing in this feeling of compassion for yourself. Now look at the image of the food again. What has changed? What are you aware of? You can experience the craving without acting on it. Just notice it, breathe, and envelope it with a sense of loving-kindness.

You can record yourself reading this script (using, for example, the Voice Memo app on the iPhone) and listen to it whenever a craving arises. You can also download an audio version of this script from our website.

TUNE IN TO YOUR BODY'S SIGNALS OF HUNGER AND FULLNESS

By mindfully tuning in to your body's cues of hunger and fullness, you may be able to reduce overeating. When you pay attention to your level of physical hunger, you are less likely to confuse it with psychological hunger. Stress, boredom, and emotions (even happy ones) can make you feel as if you're hungry even when you're really not. In a small pilot study led by psychology researcher Jennifer Daubenmier at UCSF, we found that when women are trained to do a mindful check-in before meals, they have lower blood glucose and cortisol, particularly if they are obese. And the more they improve on mental and metabolic health, the more their telomerase increases.[22] In a larger trial, psychology researcher Ashley Mason found that the more men and women practiced mindful eating, the fewer sweets they ate, and the lower their glucose was one year later.[23] Mindful eating seems to have a small effect on weight but may be critical to breaking the craving sweets–glucose link.

Below are some mindful eating strategies that I (Elissa) and my colleagues use for our studies of weight management. They are based on Mindfulness-Based Eating Awareness Training, a program developed by Jean Kristeller, a psychologist at Indiana University. (See more resources on mindful eating.)[24]

1. Breathe. Bring your awareness to your entire body. Ask yourself: How physically hungry am I right now? What information and sensations help me answer this question?

2. Rate your physical hunger on this scale:

Not at all hungry				Moderately hungry				Very hungry	
1	2	3	4	5	6	7	8	9	10

Try to eat *before* you get to 8 so you're less likely to overeat. Definitely don't wait until you're at 10. If you're famished, it's easy to eat too much, too fast.

3. When you do eat, fully savor the taste of the food and the experience of eating.

4. Pay attention to the hunger in your stomach, to the physical sensations of fullness and distention. (We call this "listening to the stretch receptors.") After you've spent a few minutes eating, ask yourself "How physically full do I feel?" Rate your answer:

Not at all full				Moderately full					Very full
1	2	3	4	5	6	7	8	9	10

Stop when your score is 7 or 8—in other words, when you're moderately full. Your biological signals of fullness, caused by increases in blood sugar and satiety hormones in the blood, kick in slowly, and you won't feel their full effect until twenty minutes later. Stopping before you get those signals, before you've eaten too much, is usually the hard part, but this becomes much easier once you start paying attention.

CHAPTER TEN

Food and Telomeres: Eating for Optimal Cell Health

Some foods and supplements are healthy for your telomeres, and some just aren't. We are happy to report that you do not need to give up carbs or milk products to be healthy! A whole-foods diet that features fresh vegetables, fruits, whole grains, nuts, legumes, and omega-3 fatty acids is not only good for your telomeres, it also helps reduce oxidative stress, inflammation, and insulin resistance—factors that, as we'll explain here, can shorten your healthspan.

It happens every single day: Morning arrives. I (Liz) am not a morning person, but I get out of bed and stagger to the kitchen, slowly waking up as I go. My husband, John, who is an early bird by nature, has kindly brewed me a cup of coffee.

"Milk?" he asks.

Well, that's a tough question for the predawn hours, made tougher by nutrition advice that often feels confusing. Yes, I like milk in my coffee. But should I pour it in? Milk is healthy, right? After all, it contains calcium and protein and is fortified with vitamin D. But should I reach for whole milk or skim? Or should I take a pass altogether?

Each additional breakfast item poses its own set of nutrition dilemmas:

Toast. Too many carbs, even if it's whole wheat? What about a potential reaction to gluten?

Butter. Will a little fat increase feelings of fullness, which is good, or will it clog the arteries, which is bad?

Fruit. Better to just ditch the toast idea and make a smoothie instead? Or...is fruit dangerously loaded with sugars?

These are a lot of questions to answer when you're still barely awake and the coffee hasn't kicked in yet. We are both scientists, trained in sifting through complicated evidence, but sometimes we still struggle to figure out what is healthiest to eat.

On mornings like this, telomeres offer a fundamental guide to the foods that are best for us. We trust telomere evidence because it looks at how the body responds to foods at the microlevel. And we take the evidence seriously, because it aligns well with the emerging knowledge in nutrition science. These findings tell us that diets don't work, and that the most empowering choice we can make is to eat fresh, whole foods instead of processed ones. As it turns out, eating for healthy telomeres is very pleasant, satisfying, and nonrestrictive.

THREE CELLULAR ENEMIES AND HOW TO STOP FEEDING THEM

You've heard us issue warnings about inflammation, insulin resistance, and oxidative stress, which create an environment that is toxic for telomeres and cells. Think of these conditions as three enemies that lurk inside each of us. You can eat foods that feed these three villains—or you can eat foods that fight them, shifting the cell environment to one that is healthier for telomere upkeep.

The First Cellular Enemy: Inflammation

Inflammation and telomere damage share a mutually destructive relationship. One makes the other worse. As we've explained, aging cells, with their short or damaged telomeres (plus any other breaks in the DNA that do not get repaired), send out proinflammatory signals that cause the body's immune system to turn on itself, damaging tissues all over the body. Inflammation can also cause immune cells to divide and replicate, which shortens telomeres even more. Thus a vicious cycle is set up.

Here's what can happen to an inflamed mouse: Researchers took a group of mice and knocked out part of a gene that protects against inflammation; without that part of the genetic code, the mice quickly developed a serious case of chronic inflammation. Their tissues accumulated short telomeres and senescent cells. The more senescent cells in their liver and intestines, the faster the mice died.[1]

One of the best ways for you to protect yourself against inflammation is to stop feeding it. The glucose absorbed from French fries or from refined carbohydrates (white bread, white rice, pasta), and from sugary candies, sodas, juices, and most baked goods, hits your bloodstream fast and hard. That uptick of blood glucose also causes an increase in cytokines, which are inflammatory messengers.

Alcohol acts as a kind of carb as well, and too much alcohol consumption appears to increase C-reactive protein (CRP), a substance that is produced in the liver and rises when there is more inflammation in the body.[2] Alcohol is also converted to a chemical (acetaldehyde, which is a carcinogen) that can damage DNA and in high doses could also harm telomeres. At least, it harms telomeres in cells in the laboratory—we have no idea if such high doses are ever achieved in humans. So far, it appears that chronic heavy alcohol use may be associated with shorter telomeres and other signs of an aged immune system, but there are no consistent relationships between light alcohol intake and telomeres.[3] It is okay to enjoy your occasional drink!

There is more good news, too, especially if you are concerned about those mice who were genetically engineered for chronic inflammation. When the mice were given an anti-inflammatory or an antioxidant drug, the telomere dysfunction was reversed. The mice's telomeres rebounded, and senescent cells stopped accumulating, so that cells could continue their dividing and renewing. This suggests that all of us can protect our telomeres from inflammation, but it's safest and smartest to do it without drugs. For a start, we can simply eat the foods that help prevent an inflammatory response from happening in the first place. And what a marvelous selection of sweet and savory plant foods we have to choose from: think of red, purple, and blue berries; red and purple grapes; apples; kale; broccoli; yellow onions; juicy red tomatoes; and green scallions. All these foods contain flavonoids and/or carotenoids, a broad class of chemicals that gives plants pigment. They are also especially high in anthocyanins and flavonols, subclasses of flavonoids that are related to lower levels of inflammation and oxidative stress.[4]

Other anti-inflammatory foods include oily fish, nuts, flaxseed, flax oil, and leafy vegetables—because all these items are rich in omega-3 fatty acids. Your body requires omega-3s to reduce inflammation and keep telomeres healthy. Omega-3s help form cell membranes throughout the body, keeping the cell structure fluid and stable. In addition, the cell can convert omega-3s into hormones that regulate inflammation and blood clotting; they help determine whether artery walls are rigid or relaxed.

It's been known for a while that people with higher blood levels of omega-3s have lower cardiovascular risk. Newer research suggests an exciting additional possibility: omega-3s may be helping to do that by keeping your telomeres from declining too quickly. Remember, telomeres shorten with age; the goal is for this shortening process to happen as slowly as possible. One study looked at the blood cells of 608 people, all of whom were middle-aged and had stable heart disease. The more omega-3s in their blood cells, the less their telomeres declined over the next five years.[5] And the less the telomeres declined, the more likely it was that these subjects, who were not so healthy to begin with, would survive the next four years.[6] Of those who had telomere shortening, 39 percent died—whereas of those who had apparent lengthening, only 12 percent died. The less your telomere length declines, the less likely *you* are to fall into the diseasespan and early death.

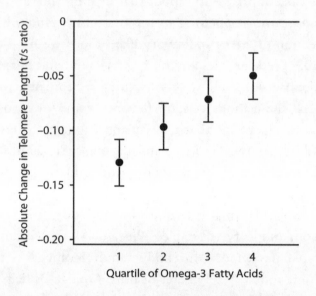

Figure 23: Omega-3 Fatty Acids and Telomere Length over Time. The higher the levels of omega-3s in the blood (EPA and DHA), the less telomere shortening over the next five years. Each one standard deviation above the average omega-3 levels predicted 32 percent lower chances of shortening. This effect was even stronger in those who started off with longer telomeres (since longer telomeres shorten more quickly).[7]

So enjoy fresh oily fish (including sushi), salmon and tuna, leafy vegetables, and flax oil and flaxseeds. (For state-by-state U.S. recommendations for fish that are caught or farmed in ways that cause less harm to the environment, consult the Monterey Bay Aquarium Seafood Watch website at http://www.seafoodwatch.org/seafood -recommendations/consumer-guides.) But should you take omega-3 supplements, otherwise known as fish-oil capsules? There has been only one randomized trial on omega-3 supplementation and telomeres, a study by psychologist Janice Kiecolt Glaser at Ohio State, and the results were suggestive. She found that people who took fish-oil supplements for four months did not have longer telomeres than people who took a placebo. However, across all the groups, the greater the increases in omega-3s in the blood relative to their levels of omega-6 fatty acids, the greater the telomere lengthening over that period.[8] The omega supplementation also reduced inflammation, and the greater decreases in inflammation were associated with increases in telomere length. (Omega-6s are polyunsaturated fats that come from sources like corn oil, soybean oil, sunflower oil, seeds, and certain nuts.) We must note, though, that those taking the supplements had other significant telomere-friendly changes: reduced levels of oxidative stress and inflammation. The results appear to depend on how well each person absorbed the levels of omega-3 polyunsaturated fats from the supplements.

Your blood levels of omega-3s, or any nutrient, are not always directly related to whether you're consuming dietary or supplement sources. All kinds of complicated and mostly unknowable factors affect this number: how well you absorb the nutrient, how well your cells use it, how fast you metabolize and lose it. (This is a good piece of information to keep in mind whenever you read recommendations for diet and supplementation.) In general, we suggest that everyone try to get their nutrients from their diet, but when that's just not possible, supplementation can be a reasonable alternative (make sure you check with your physician first). Even the most innocent-seeming supplements can have side effects or interact with medications you're already taking. They could also be contraindicated for people with certain health conditions. A general consensus seem to be a daily dosage of at least 1,000 milligrams of a mixture of EPA and DHA, which is similar to the low dose tested in the Ohio State study. For sustainability reasons, we strongly suggest the vegetarian alternative, which is made from algae. Fish have omega-3s because they eat algae. We can eat algae, too, sustainably farmed algae that contains DHA. The oceans cannot support enough fish oil to maintain the world's healthy telomeres. So far, it appears the DHA from algae promotes similar benefits to cardiovascular health as the DHA from fish.

Telomere research suggests that you should make consumption of omega-3s a priority. But you also have to keep an eye on the balance between your omega-3s and omega-6s, because the typical Western diet tilts us more toward omega-6s than omega-3s. To keep your omegas in balance, we suggest that you keep eating healthy, unprocessed foods like nuts and seeds—but dramatically reduce your consumption of fried foods, packaged crackers, cookies, chips, and snacks, which often contain oils made with high amounts of omega-6s, as well as saturated fats, which are a risk factor for cardio-vascular disease.

There's another chemical in our body worth getting to know: homocysteine, which is chemically related to cysteine, one of the amino acid building blocks of proteins. Homocysteine levels go up with aging, and correlate with inflammation, wreaking havoc on the lining of our cardiovascular system to promote heart disease. In many studies, having high homocysteine is associated with having short telomeres. But telomeres reflect the input of many factors. So it is no surprise that in one study, the relationship between telomeres and mortality appears to be due in part to both high inflammation and high homocysteine—we don't know which came first.[9] The good news here is that if you have especially high homocysteine, this is one of the cases where a vitamin pill might help—B vitamins (folate or B_{12}) appear to reduce homocysteine.[10] (Check with your doctor to see if you should be taking this supplement.)

The Second Enemy: Oxidative Stress

Human telomeres have a DNA sequence that looks like this: TTAGGG, repeated in tandem over and over, commonly over a thousand times at each chromosome end. Oxidative stress—that dangerous condition that occurs when there are too many free radicals and not enough antioxidants in your cells—damages this precious sequence, especially its GGG segments. Free radicals take aim at that big juicy row of GGG pieces, a particularly sensitive target. After free radicals have their way, the DNA strand is broken; the telomere gets shorter faster.[11] It's as if the rich meal of GGGs has been fed to the cellular enemy, oxidative stress. In cells grown in the lab, oxidative stress damages telomeres, and it also reduces the telomerase activity that can replenish shorter telomeres. It's a double whammy.[12]

But if you pump up the cells' medium (the liquid soup that supports a cell's life when it's sitting in a lab flask) with vitamin C, the telomere is protected from the free radicals.[13] Vitamin C and other antioxidants (like vitamin E) are scavengers that gobble up free radicals, preventing them from harming your telomeres and cells. People with higher blood levels of vitamins C and E have longer telomeres, but only when they also have lower levels of a molecule known as F2-isoprostane, which is an indicator of oxidative stress. The higher this ratio between blood antioxidants and F2-isoprostane, the less oxidative stress there is in the body. This is just one of the many reasons you should eat fruits and vegetables every day; they offer some of the best sources of antioxidant protection. To get sufficient antioxidants in your diet, eat plenty of produce, especially citrus, berries, apples, plums, carrots, green leafy vegetables, tomatoes, and, in smaller portions, potatoes (red or white, with the skin on). Other plant-based sources of antioxidants are beans, nuts, seeds, whole grains, and green tea.

At this point, we don't suggest getting your antioxidants from a supplement if your goal is telomere health. That's because the evidence for a connection between antioxidant supplements and healthy telomeres is still inconclusive. Some studies have found that the higher the level of certain vitamins in the blood, the longer the telomeres, and we have listed these in the table on page 238. However, while some studies find multivitamin use accompanies longer telomeres,[14] at least one study found that taking a multivitamin was related to shorter telomeres.[15] Also, high antioxidant levels even provoked laboratory-grown human cells to take on certain cancerous properties, a finding that, again, may warn us that too much of a good thing may be, simply, too much. In general, antioxidants from food are typically better absorbed by the body and may have more powerful effects than supplements.

OUR EARLIEST NUTRITION

Can you feed your baby's telomeres? Possibly, by making sure your baby is exclusively breast-fed in its first weeks. Janet Wojcicki, a health researcher at UCSF who has been following cohorts of pregnant women, found that children who were breast-fed only (no formula or solid foods) in the first six weeks of life have longer telomeres. Solid foods can cause inflammation and oxidative stress when they are given to infants whose guts are not yet ready.[16] Perhaps that is why introducing solid food before six weeks of age is linked to shorter telomeres.

The Third Enemy: Insulin Resistance

Nikki, a physician and administrator at her hometown hospital, has a vice: massive consumption of the sugared soda Mountain Dew. She developed the habit in residency, when she learned to rely on its sugar and caffeine to stay awake. The habit has remained with her. Early each morning, Nikki pulls a one-liter bottle of Mountain Dew from a small refrigerator in her garage, which is dedicated to warehousing her stash. She sets the bottle in the passenger seat of her car on the way to work. At each stoplight, she unscrews the bottle and takes a swig. When she arrives at work, the bottle goes into the fridge. After grand rounds: a swig. After a meeting: swig. After finishing some paperwork: swig. By the end of her long, grueling day, the bottle is empty. "I couldn't get through without it," Nikki says, with a fatalistic shrug of her shoulders.

As a doctor, Nikki knows that a daily one-liter dose of Mountain Dew is not a healthy habit. But like nearly half of all Americans, she drinks soda anyway. These folks might as well give the third enemy—insulin resistance—a straw and say, "Drink up; this stuff will help you get as big and as terrifying as you want to be."

Here's a frame-by-frame shot of what happens when you swallow sugary soda, or "liquid candy": Almost instantaneously, the pancreas releases more insulin, to help the glucose (sugar) enter cells. Within twenty minutes, glucose has built up in the bloodstream and you have high blood sugar. The liver starts to turn sugar into fat. In about sixty minutes, your blood sugar falls, and you start thinking about having more sugar to pick you back up after the "crash." When this happens often enough, you can end up with insulin resistance.

Is soda the new smoking? Maybe. Cindy Leung, a nutritional epidemiologist at UCSF and one of our collaborators, found that people who drink twenty ounces of sugary soda daily have the equivalent of 4.6 extra years of biological aging, as measured by telomere shortness.[17] That, astonishingly, is about the same level of telomere shortness caused by smoking. When people drink eight ounces of soda, their telomeres are the equivalent of two years older. You may be wondering if people who drink soda have other unhealthy habits that may affect the results—and that's a great question. In this study, which looked at around five thousand people, we did what we could to address confounding factors. We corrected for some available factors, including diet and smoking; and then we corrected for all available factors, including diet, smoking, BMI, waist circumference (to gauge belly fat), income, and age, that might have otherwise explained away this association. The association did not go away. This association between soda and telomeres exists in young children, too. Janet Wojcicki found that at three years old, children who were drinking four or more sodas a week had a greater rate of telomere shortening.[18]

Sports drinks and sweetened coffee drinks are liquid candy,

too. They contain as much sugar as a typical soda (42 grams in a 12-ounce-tall Peppermint Mocha from Starbucks) so it is wise to stay away from them, or to drink them only rarely, as a special treat.[19] Sodas and sweetened beverages are a dramatic example of sugar's harm to telomeres, because of the delivery method. It's a fast rush of sugar with no fiber to slow it down. Almost anything that's considered a dessert or a treat is a source of high sugar: cookies, candy, cakes, ice cream. Once again, refined products like white bread, white rice, pasta, and French fries are high in simple or rapidly absorbed carbohydrates and can wreak havoc on your blood sugar levels, too.

To prevent insulin spikes that can eventually lead to insulin resistance, focus on foods that are higher in fiber: Whole-wheat bread, whole-wheat pasta, brown rice, barley, seeds, vegetables, and fruits

Figure 24: Finding a Balance—as Guided by Telomeres. Choose more foods high in fiber, antioxidants, and flavenoids, like fruits and vegetables. Include foods high in omega-3 oils, like seaweed and fish. Choose less refined sugars and red meat. A healthy dietary balance, like the one pictured above, will lead to healthy shifts in your blood to high nutrients and less oxidative stress, inflammation, and insulin resistance.

are all excellent sources. (Fruits, although they contain simple carbo-hydrates, are healthy because of their fiber content and overall nutritional value; fruit juices, from which the fiber has been extracted, are generally not.) These foods are also filling, which helps you avoid eating excess calories. They are the same foods that help reduce the belly fat that is so closely associated with insulin resistance and metabolic disorder.

VITAMIN D AND TELOMERASE

Higher levels of vitamin D in the blood predict lower overall mortality rates.[20] Some studies find that vitamin D is related to longer telomere length, more so in women than men, and other studies do not find a relationship. So far we've found one study that tested the effects of supplements: In that small study, 2,000 IU a day of vitamin D (in the form of vitamin D$_3$) for four months led to increased telomerase by around 20 percent compared to a placebo group.[21] While the jury is still out on a relationship with telomeres, it's notable that vitamin D levels are often low, depending on where you live and on sunlight exposure. The best dietary sources of vitamin D are salmon, tuna, sole, flounder, fortified milk and cereals, and eggs. It can be hard to get enough vitamin D from diet and sunlight alone, depending on where you live, so this is a case when you may want to consider supplements (consult your doctor).

A HEALTHY EATING PATTERN

Platters of freshly caught fish, bowls heaped with fruits and vegetables in deep, rich hues, dishes of hearty beans, whole grains, nuts and seeds...it's a menu for a feast. It's also a recipe for supporting a healthy cellular environment. These foods reduce inflammation, oxidative stress, and insulin resistance. These foods fit into a healthy eating pattern that is great for telomeres and overall health.

Around the world—from Europe to Asia to the Americas—eating habits can be very roughly divided into two categories. There are people whose diets feature lots of refined carbohydrates, sweetened sodas, processed meat, and red meat. And then there are people who have a high intake of vegetables, fruits, whole grains, legumes, and low-fat, high-quality sources of protein, including seafood. This healthier diet is sometimes called the Mediterranean diet, but most cultures around the globe have some version of this eating pattern. Some of the details vary—some cultures eat more dairy or seaweed—but the general idea is to eat a variety of fresh, whole foods, and for most of these foods to come from a low spot on the food chain. Some researchers call this the "prudent dietary pattern." That's an accurate label, though it doesn't capture just how delicious and healthy these foods are.

People who follow this prudent pattern have longer telomeres, no matter where they live. In Southern Italy, for example, elderly people who followed the Mediterranean diet had longer telomeres. The more closely the adhered to this type of a diet, the better their overall health and the more they could fully participate in the activities of daily living.[22] And in a population study of middle-aged and older people in Korea, people who followed the local version of a prudent dietary pattern (i.e., more seaweed and fish) had longer telomeres ten years later than people who ate a diet high in red meat and refined, processed foods.[23]

We've been speaking of broad dietary patterns, but what are the best particular foods for healthy telomeres? The Korean study gives us a clue. The more that people ate legumes, nuts, seaweed, fruits, and dairy products, and the less they consumed red meat or processed meat and sweetened sodas, the longer their telomeres in their white blood cells.[24]

The benefits of eating wholesome, unprocessed foods—and not too much red meat or processed meats—hold strong across the world, through adulthood, and all the way into old age. In 2015, the World Health Organization identified red meat as a probable cause of cancer and processed meat as a cause.[25] When types of meat are examined

236

in telomere studies, processed meat appears worse for telomeres than unprocessed red meat.[26] Processed meat refers to meat that has been altered (smoked, salted, cured), such as hot dogs, ham, sausage, or corned beef.

Of course, it is best to eat well throughout your entire life, but it is never too late to begin. The chart that follows can help guide your daily food choices. In general, though, we suggest that you worry less about any particular food item (an attitude that makes mornings easier for Liz) and focus instead on eating a variety of fresh, wholesome foods. You'll find yourself enjoying foods that fight inflammation, oxidative stress, and insulin resistance, without needing to plan carefully in advance. And you will find that you naturally follow the kind of eating plan that is healthy for your telomeres. Plus, you won't shorten your telomeres by worrying too much about all of the food choices you make every day!

BEANS ABOUT COFFEE?

Coffee's effects on health have been questioned in hundreds of studies. Those of us who love our morning cup will be happy to hear that it almost always turns up innocent. Meta-analyses show coffee reduces the risk of cognitive decline, liver disease, and melanoma, for example. Only one trial has been done on coffee and telomere length, but the news so far is good: Researchers tested whether coffee might improve the health of forty people with chronic liver disease. They were randomized to either drink four cups of coffee a day for a month or refrain (the latter were the controls). After the period of drinking coffee, the patients had significantly longer telomeres and lowered oxidative stress in their blood than the control group.[27] Further, in a sample of over four thousand women, those who drank caffeinated coffee (but not decaffeinated) were likely to have longer telomeres.[28] More reasons to enjoy the aroma of your morning coffee brewing.

We have discussed vitamin D and omega-3 supplements, which are often found to be deficient. However, outside of these, we do not make specific recommendations on supplements, because each person's needs are different, and nutrition studies' conclusions about supplements are notoriously changed by new studies. It's hard to be confident about the effects and safety of high doses of anything.

NUTRITION AND TELOMERE LENGTH**

Food, Drinks, and Telomere Length	
Associated with Shorter Telomeres	**Associated with Longer Telomeres**
Red meat, processed meat[29] White bread[30] Sweetened drinks[31] Sweetened soda[32] Saturated fat[33] Omega-6 polyunsaturated fats (linoleic acid)[34] High alcohol consumption (more than 4 drinks per day)[35]	Fiber (whole grains)[36] Vegetables[37] Nuts, legumes[38] Seaweed[39] Fruits[40] Omega-3s (e.g., salmon, arctic char, mackerel, tuna, or sardines)[41] Dietary antioxidants, including fruits, vegetables, but also beans, nuts, seeds, whole grains, and green tea[42] Coffee[43]
Vitamins	
Associated with Shorter Telomeres	**Associated with Longer Telomeres**
Iron-only supplements[44] (probably because they tend to be high doses)	Vitamin D[45] (mixed evidence) Vitamin B (folate), C, and E Multivitamin supplements (mixed evidence)[46][47]

**Note that the scientific literature here is growing and changing all the time. Check our website for updates!

TELOMERE TIPS

- Inflammation, insulin resistance, and oxidative stress are your enemies. To fight them, follow what's been called a "prudent" pattern of eating: Eat plenty of fruits, vegetables, whole grains, beans, legumes, nuts, and seeds, along with low-fat, high-quality sources of protein. This pattern is also known as the Mediterranean diet.
- Consume sources of omega-3s: salmon and tuna, leafy vegetables, and flax oil and flaxseeds. Consider supplementation with an algae-based omega-3 supplement.
- Minimize red meat (especially processed meat). You might try to go vegetarian for at least a few days each week. Eliminating meat can benefit your cells as well as the environment.
- Avoid sugary foods and drinks, and processed foods.

RENEWAL LAB

TELOMERE-FRIENDLY SNACKS

Healthy snacks are important to have on hand, because the alternative is usually *unhealthy* snacks. Typical snack foods are often processed and contain unhealthy fats, sugars, and salts. We recommend any whole-food snack high in protein and low in sugars. Here are a few ideas that also include high levels of either antioxidants or omega-3 polyunsaturated fats.

Homemade trail mix. Making your own trail mix is easy, and it's the best way to make sure it is low in sugar. (Store-bought trail mix often hides added sugars in dried fruits.) This mix is high in omega-3s and antioxidants. It's also rich in energy, so be sure to enjoy it in moderate quantities.

Combine:

- 1 cup walnuts
- ½ cup cacao nibs or dark chocolate chips
- ½ cup goji berries or other dried berries

Optional additions:

- ½ cup dried unsweetened coconut flakes
- ½ cup raw or unsalted sunflower seeds
- 1 cup raw almonds

Homemade chia pudding. Chia seeds are high in antioxidants, calcium, and fiber. These unassuming little seeds from South America also house 28 grams of omega-3s in every ounce. Chia pudding is a great snack, but it makes a tasty part of breakfast, too.

Combine:

- ¼ cup chia seeds
- 1 cup unsweetened almond or coconut milk
- ⅛ teaspoon cinnamon
- ½ teaspoon vanilla extract

After stirring the ingredients together, let the mixture sit for five minutes. Stir the pudding again and place in the fridge for 20 minutes, or until thick, or overnight.

Optional garnishes:

- dried coconut flakes
- goji berries
- cacao nibs
- sliced apple
- honey

Seaweed. Yes, seaweed. It's an easy grab, and it's telomere friendly. Seaweed snacks, such as SeaSnax, can be found in health-food stores and are made from seaweed sheets lightly roasted in olive oil with a pinch of sea salt. They come in different flavors (we especially like wasabi or onion) and are a great snack for people who crave salty or savory foods. Seaweed is also extremely rich in micronutrients, so enjoy. If you are watching your sodium, choose unsalted sheets of seaweed.

KICK A BAD FOOD HABIT: FIND YOUR MOTIVATION

Adding healthy foods into your diet is great, but it may be even more important to avoid the kind of processed, sugary, junky foods that feed your cellular enemies. Breaking an unhealthy food habit is easier said than done. When people identify their personal motivation for changing a habit, they are more likely to successfully make that change. Here are some of the questions that we ask our research volunteers to help them identify their most meaningful goals when they are trying to make changes to their diet:

- *How is your diet affecting you? Has anyone ever encouraged you to cut down on something? Why? What do you most want to change?*
- *Why exactly are you concerned about how much fast food (or junk food, sugar, or other unhealthy food) you eat? Do you have diabetes or heart disease in your family history? Do you want to lose weight? Are you worried about your telomeres?*
- *What part of you wants to change? What part of you doesn't? What are the things you care most about? How would making this change impact you and people you care about?*

When you've identified your strongest source of motivation, visualize it. If your motivation is to live a long, healthy life, create a vivid image of yourself being active and healthy at age ninety, or cheering at your grandchild's graduation. Do you want to make sure you're around to see your children grow up? Picture yourself dancing at their wedding receptions. Perhaps thinking of those tiny telomeres bravely protecting the future of your chromosomes in billions of cells throughout your body will motivate you! Whenever you're facing temptation, call that image to mind. Our colleague Professor Len Epstein of SUNY Buffalo has found that thinking vividly about the future helps people resist overeating and other impulsive behaviors.[48]

MASTER TIPS FOR RENEWAL:
Science-Based Suggestions for Making Changes That Last

Behavior change is simple, and behavior change is hard. For some people, learning about telomeres is a potent motivator. They imagine their telomeres eroding away—and they are spurred to get more exercise, for example, or adopt a challenge response to stress.

Often, though, motivation is not enough.

The science of behavior change tells us that if you want to make a change, you need to know why you're making the change—but for that change to really last, you need *more* than knowledge. When it comes to change, our minds don't work rationally. We operate largely out of automatic patterns and impulses. Thus the donut instead of the vegetable omelet, the firm resolutions that weaken when it's time to work out or meditate. As a species, we humans have much less personal control than we would like to think. Fortunately, behavioral science tells us how to make changes that stick.

First, identify a change you'd like to make. **The self-test (Telomere Trajectory Quiz) starting on page 162 can help you see where your telomeres need the most help.** Choose one area (such as exercise) and a change you'd like to make (such as starting a walking program). Before you make that change, ask yourself three questions:

1. On a scale of 1 to 10, how do you rate your readiness

243

to make this change? (A ranking of 1 would mean you're not at all ready; 10 means extremely ready.) If you rank your readiness at 6 or lower, go to the question below to explore what truly motivates you. Then, rate your readiness again. If your readiness score does not increase, choose a different goal.

Many of us engage in behaviors we'd like to change, but we feel stuck or ambivalent. Find one small behavior you feel ready to focus on now. One change leads to another, so focusing on one small change is the right place to put your efforts now. For tough, compulsive behaviors like excessive smoking, drinking, and overeating, you might consult a professional coach or therapist who is an expert in "motivational interviewing," a dialogue that helps people develop clear goals, get past obstacles, and meet the goals.[1]

2. What about this change is meaningful to you? Ask yourself what things are most important to you. Try to tie your goal to your deepest priorities in life, as in "I want to begin a walking program, because I want to be healthy and independent, in my own home, for as long as possible." Or "I want to be active in the lives of my child and grandchildren." The tighter the connection between your goal and your values and priorities, the more likely you are to stick with the change. Choosing intrinsic goals—those related to relationships, enjoyment, and meaning in life—works better than choosing external goals (which tend to be about wealth, fame, or how others view us). They have more lasting power for behavior change and bring us more happiness.[2]

Ask yourself the hard questions in the Renewal Lab for Chapter 10 (page 242) about finding your motivation. Then take a mental snapshot of the answer, an image that represents your motivation. This visual picture is a weapon you can use in those difficult moments when part of you is looking hard for a way out of the new behavior.

3. On a scale of 1 to 10, how confident are you that you can make this change? If you're at a 6 or lower, change your goal to make it smaller and easier to achieve. Identify any obstacles that dragged your rating down and make a realistic plan for overcoming them. Think of obstacles with that "challenge" mind-set—this is an opportunity to bring in some good stress. Another way to increase efficacy and success is to think of a past proud moment when you overcame an obstacle.[3]

Self-efficacy ratings like this are our crystal ball; they've been shown to be one of the biggest predictors of our future behavior. Confidence about whether we can carry out a specific task determines a cascade of events: whether we will even try a new behavior in the first place, whether we will persist at it once we hit obstacles.[4] Get into the self-efficacy positive loop—achieving a small part of our goal boosts our confidence, which carries us to the next step, which boosts our confidence further.

Next, consider whether you're trying to create a new habit or to break an old one. The answer will determine which strategies apply to you.

TIPS FOR CREATING NEW HABITS

Our brains are equipped for automaticity, for making the least effort possible. Make automaticity work for you, not against you. Here's how:

- **Small changes.** Slip into your new habit painlessly, in small doses. If you want to get more sleep, don't try to go to bed an hour earlier each night. That's too hard. Start by going to bed fifteen minutes earlier each night. If that's not doable, start with an even smaller goal: ten minutes, five minutes...whatever feels easy and nonthreatening. From there, you can build slowly toward your goal.
- **Staple it.** Tack your small change onto an activity that's already a routine part of your day.[5] That way, you'll have to think less about when to make the change, and eventually it will become

245

routine, too. For example, whenever I (Liz) wait for my computer to complete loading my e-mail, it's a trigger for me to do a micro-meditation. For other people, the lunch break is a trigger that it's time to take a walk. Hitching the behavior to an already embedded one helps you stick to your plan.

- **Mornings are green light zones.** Try to schedule your change for the morning. The earlier in the day, the less likely it is that other urgent priorities will nudge your new behavior off your schedule. You may feel stronger determination, which you can visualize as a green light that flashes "DO IT."

- **Don't decide—just do.** When it's time to go to the gym (or make any other change), don't ask yourself "Should I?" Making decisions is exhausting. And in a weak moment the answer may well be "Tomorrow." Just go. Walk there like an unthinking zombie if you have to.

- **Celebrate it.** Have a quick mini-celebration each time you practice your new habit. Consciously say to yourself, "Great!" or "I did it!" or "DONE!" and let yourself feel pride. Or put aside a dollar each time in a collection for some personal indulgence after ten times.

TIPS FOR BREAKING OLD HABITS

Trying to end an old, unwanted habit takes willpower, which is, sadly, a limited resource. Plus, plenty of unhealthy habits make us feel good, at least for a few moments. Sugary foods and drinks, for example, make your brain's reward system light up. We can become neurobiologically dependent on that sugar rush. Breaking the habit takes patience and persistence.

- **Increase your brain's ability to execute your plans.** We are best able to exert control when the brain networks that foster analytical thought are activated. When there is more activity in the prefrontal cortex, some of the more emotional areas in the amygdala are inhibited. Exercise, relaxation meditation, and foods that are high in quality protein promote this optimal mental state (and stress thwarts it).

- **Don't try the change when you are feeling depleted.** Sleep loss, low blood sugar, or high emotional stress can deplete you of your willpower. Wait until the conditions are in your favor.[6]

- **Shape your surroundings to reduce the number of times you're tempted.** Don't keep sweets, soda, or other reminders of your unwanted habit around the house, and certainly not within sight. Cookies and chips, when they do end up at home, should be out of sight in a high cupboard, not on the kitchen counter in a bowl. You may be able to resist temptation once, but saying no several times a day is exhausting. Your limited supply of willpower may run out. These tips are called stimulus control—we attempt control over our environment as much as we can so we are not surrounded by the tempting stimuli.

- **Follow your natural alertness rhythms.** You'll have more energy to stoke your willpower. If you're a night owl, you'll be more able to resist temptation in the evening, and more likely to succumb in the early morning. Plan accordingly. And take healthy snack breaks at your personal low points— the times of day when you tend to feel tired. This will sustain your energy for when you need to draw on your willpower.

Last, there is one strategy that helps almost everyone in every case, regardless of whether you are trying to start or stop something: social support. Ask your family and friends to help support your new goal. Tell them what would be helpful. Turn your accomplices (those who help you do exactly what you don't want to do) into supportive influences or... avoid them! You could find a partner with similar goals to share the journey with you. I (Elissa) would go running less often if I didn't have a running partner relying on me.

To help you think about ways to make small changes throughout your day, we've created the reference "Your Renewed Day" on the next page. It's a time-based table that shows you which routine behaviors can endanger your telomeres. It also suggests substitute actions that are telomere healthy.

Your Renewed Day

Each day you have an opportunity to forestall, maintain, or accelerate the aging of your cells. You can stay in balance or maybe even forestall unnecessary acceleration of biological aging by eating well, getting enough sleep for restoration, being active and maintaining or building fitness, and sustaining yourself through meaningful work, helping others, and social connection.

Or you can do the opposite—consume junk food or too many sweets, get too little sleep, and stay sedentary or decondition the fitness you have. Throw high stress into the mix of a vulnerable body, and you'd have a day of wear and tear on your cells. It's possible that you might even lose a few base pairs of telomere length. We don't *really* know how responsive telomeres are on a daily basis, but we do know that chronic behavior over time has important effects. We can all strive to have more days of renewal rather than wear and tear. Begin by making small changes. There are suggestions for telomere-healthy change throughout the book, and we've created an example of how you can build some of these behaviors into your day. Circle any you might like to try.

We've also included a blank Renewed Day schedule that you can customize with the telomere-healthy changes you'd like to make. You can copy it, or print it from our website, and stick it to your refrigerator or mirror to help remind you of easy ways to promote healthy cell renewal. Fill in several new behaviors you'd like to add to your day. What do you want to say to yourself when you wake up? Do you want to fit in a few minutes of a morning renewal

Your Renewed Day

Time	Telomere-Shortening Behavior	Telomere-Supporting Behavior
Waking up	Anticipatory stress or dread.	**Reappraise your stress response** (page 120). Wake with joy. "I am alive!"
	Mentally rehearse your to-do list. Check phone immediately.	Set an intention for the day. Look forward to any positive aspects.
Early morning	Regret that there's no time for exercise.	Perform a **cardio or interval workout** (page 186). Or do energizing **Qigong** (page 156)
Breakfast	Sausage and bagel.	Oatmeal with fruit; fruit smoothie with yogurt and nut butter; vegetable omelet.
Morning commute	Rush, hostile thoughts, maybe a little road rage.	Practice the **three-minute breathing break** (page 149).
Arrival at work	Play catch-up from the moment you arrive.	Give yourself a ten-minute window of habituation and settling before work begins.
	Anticipate, worry about the workday.	Meet situations as they arise.
Workday	Self-critical thoughts.	Notice your thoughts. Take a **self-compassion break** (page 122) or **manage your eager assistant** (page 123).
	Multitask to deal with work overload.	Focus on one task at a time. (Can you turn off your e-mail and ringer for an hour?)
Lunch	Eat fast food, deli meats.	Enjoy a lunch made from fresh, whole foods.
	Eat quickly.	Practice **mindful eating** (page 222).
		Connect with someone. Have lunch or walk with a partner; text, call, or e-mail someone you have a supportive relationship with.
Afternoon	Give in to cravings for a sugared drink, baked goods, or candy.	**Surf the urge** (page 220). Have a **telomere-friendly snack** (page 240).
		Stretch.
Evening Commute	Ruminate.	**Mentally distance yourself** (page 97).
	Negative mind wandering.	Take a **three-minute breathing break** (page 149).
Dinner	Eat processed food.	Have a whole-foods dinner (see our website for ideas).
	Look at screens.	Give the gift of focused attention to others.
Evening	Run through your evening activities and chores without a break.	**Exercise**, or try a **stress-reducing technique** (page 153).
		Ask, "Did I live my intentions today?"
	Suffer from a head buzzing with the aftereffects of a busy nonstop day.	Review your day; try a **challenge reappraisal** (page 87). Savor the things that made you happy.
		Engage in a **relaxing sleep ritual** (page 203).

My Renewed Day

Waking up	
Early morning	
Breakfast	
Morning Commute or Arrival at Work	
Workday	
Lunch	
Afternoon	
Evening Commute or Arrival at Home	
Dinner	
Evening	

mind-body activity? Think about transitions in the day when you can build in more physical activity, shift your awareness to the moment to promote stress resilience, connect with other people, and add some telomere-healthy foods to your diet.

Just remember that the path to lasting change is traveled one small step at a time.

PART IV

OUTSIDE IN:
THE SOCIAL WORLD
SHAPES YOUR
TELOMERES

CHAPTER ELEVEN

The Places and Faces That Support Our Telomeres

Like the thoughts we think and the food we eat, the factors beyond our skin—our relationships and the neighborhoods we live in—affect telomeres. Communities where people do not trust one another, and where they fear violence, are damaging to telomere health. But neighborhoods that feel safe and look beautiful—with leafy trees and green parks—are related to longer telomeres, no matter what the income and education level of their residents.

When I (Elissa) was a graduate student at Yale, I routinely worked late into the evening. By the time I walked back home from the psychology building, it was dark. I had to pass a church where someone had been murdered a few years earlier, and even though the area was usually quiet when I walked by at around 11:00 p.m., my heart would beat faster. Next, I turned down my street, where the rent was quite affordable on a student stipend. It was a long street, known for occasional muggings. As I walked, I listened carefully for steps behind me. I could feel my heart thump more powerfully. It is a good bet that my blood pressure went up and that glucose was recruited from its stores in my liver, giving me energy to run if needed. Every night, my body and mind mobilized themselves for danger. That experience lasted for just ten minutes each evening.

Imagine how stressful it would feel if the risk was much worse, the duration was longer, and you couldn't afford to move away.

Where we live affects our health. Neighborhoods shape our sense of safety and vigilance, which in turn affects levels of physiological stress, emotional state, and telomere length. Besides violence and lack of safety, there is another critical aspect that makes neighborhoods potent influences on health, and that is the level of "social cohesion"—the glue, the bond, among people who live in the same area. Are your neighbors mutually helpful? Do they trust one another? Do they get along and share values? If you were in need, could you rely on a neighbor?

Social cohesion is not necessarily a product of income or social class. We have friends in a beautiful gated neighborhood, where the houses sit on acres of rolling hills. There are positive signs of social cohesion, including Fourth of July picnics and holiday dances. But there's also mistrust and infighting, and it's not free of crime. It's a neighborhood full of doctors and lawyers, but if you live there you might wake up in the morning to the sound of a police helicopter hovering over your house, searching for an armed robbery suspect who has jumped over the gate. When you take out the trash, a neighbor who is unhappy about your plans for remodeling might accost you. Check your messages, and you could find that your neighbors are in a heated e-mail fight about whether to hire a security patrol and who will pay for it. You may not even know the person who lives next door. There are also neighborhoods that are poor but have people who know each other and have a strong sense of community and trust. While income plays a role, a neighborhood's health goes way beyond income.

People in neighborhoods with low social cohesion and who live in fear of crime have greater cellular aging in comparison to residents of communities that are the most trusting and safe.[1] And in a study in Detroit, Michigan, feeling stuck in your neighborhood—wanting to move but not having the money or opportunity to do so—is also linked to shorter telomeres.[2] In a study based in the Netherlands (known as the NESDA study), 93 percent of the sample rated their neighborhood as generally good (or higher). Despite that these neighborhoods were good environments, the more specific ratings of quality—including levels of vandalism and perception of safety—were associated with telomere length.

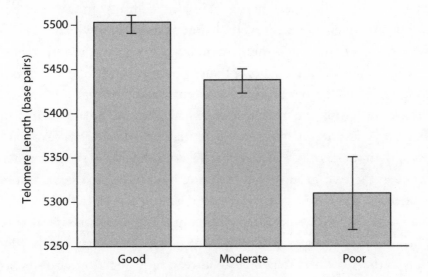

Figure 25: Telomeres and Neighborhood Quality. Here in the NESDA study, residents of neighborhoods with higher quality had significantly longer telomeres than those with moderate or poor quality.[3] This is even after adjusting for age, gender, demographic, community, clinical, and lifestyle characteristics.

Maybe the people living in lower quality neighborhoods have more depression. Was that a possibility that occurred to you? It makes sense that people who live in neighborhoods with low social cohesion would feel worse psychologically. And we know that depressed people have shorter telomeres. The NESDA researchers tested this—and found that the emotional stress of living in an unsafe neighborhood has an effect independent of how depressed or anxious its residents are.[4]

Exactly how does low social cohesion penetrate to your cells and telomeres? One answer has to do with vigilance, that sense of needing to be on high alert to maintain your safety. A group of German scientists performed a fascinating study of vigilance that pitted country folks against city dwellers. People from both groups were invited to take one of those nerve-wracking math tests that are designed to elicit a stress response, the kind where volunteers perform complex mental math while researchers give instant feedback. In this case, the participants were hooked up to a functional MRI, which allowed the researchers to watch their brain activity, and so the researchers gave their feedback over headphones, saying things like "Can you go faster?" or "Error! Please start from the beginning." When the city dwellers took the math test, they had a bigger threat response in their amygdala, a tiny brain structure that is the seat of our fear reactions, than the people who lived in the country.[5] Why the difference between the two groups? Urban living tends to be less stable, more dangerous. People in cities learn to be more vigilant; their bodies and brains are always prepared to mount a big, juicy stress response. This ultrapreparedness is adaptive but is not healthy, and it may be part of the reason that people in threatening social environments have shorter telomeres. (It's interesting, and a source of relief to us city dwellers, that the noise and crowds of urban living are *not* associated with shorter telomeres.)[6]

Some neighborhoods may shorten telomeres because they are places where it's harder to maintain good health habits. For example, people tend to get less sleep when they live in neighborhoods that are disorderly and unsafe, with low social cohesion.[7] Without adequate sleep, your telomeres suffer.

I (Liz), who also lived in New Haven for a time, experienced firsthand another way that a neighborhood can inhibit health habits. Prior to moving to New Haven, I had studied in Cambridge, England. With its flat terrain, Cambridge is a bicycling haven, and I rode my bike everywhere. When I arrived in New Haven to start postdoctoral research at Yale, I noticed that its geography was ideal for cycling. One of the first questions I asked my new lab mates was, "Where can I get a bike to ride to and from work?"

A short silence followed. Someone said, "Well, maybe biking home in the evening is not such a good idea. Bikes tend to be stolen."

Airily I replied that when that had happened in Cambridge, I simply bought a cheap, secondhand bike to replace it. Another silence, and then someone kindly explained that when their colleague said "stolen," he had meant "stolen while the person was still riding the bike." So I didn't bike in New Haven.

Other residents of low-trust, high-crime neighborhoods may draw similar conclusions. It's hard enough for many of us to fit exercise into our schedules, or to resist the call of the easy chair—and for people in unsafe neighborhoods, certain kinds of exercise may be too dangerous to even contemplate. Safety is only one barrier. Another is a lack of parks and other places to exercise. The social and "built environment" of poor neighborhoods stacks the deck against exercise. Without exercise, telomeres are shorter.

LITTERED OR LEAFY?

San Francisco is one of the world's great cities. Its citizens live within walking distance of museums, restaurants, and theaters; they can hike to spectacular views of the hillsides and bay. But as with many cities, parts of San Francisco are also quite dirty. They have a litter problem. This is not good for the residents, especially the young ones. Children who live in a neighborhood that is physically disorderly, with vacant buildings and trash in the streets, have shorter telomeres. The presence of litter or broken glass right outside the house is an especially strong predictor of telomere trouble.[8]

Have you ever been to Hong Kong? There is a stark contrast between the densely populated bustle, bright neon lights, and chaos of Kowloon, the city's downtown, and the sprawling green hills of the New Territories, which are located just outside the city. There, the citizens enjoy trees, parks, and rivers. A 2009 study looked at nine hundred elderly men; some lived in Kowloon and others lived in the lush New Territories. Guess who had shorter telomeres? The men who lived in the city. (The study controlled for social class and health behaviors.) While other factors could be responsible for the association, this study suggests that there's a role for green space in telomere health.[9]

When you're in the thick of a forest, breathing the crisp, clean air, it's not hard to believe that telomeres could benefit from exposure to nature. We're intrigued by this possibility because it's supported by what we already know about nature and a phenomenon called psychological restoration. Being in nature provides a dramatic change in context. It can inspire us with beauty and stillness. It takes us out of small thinking about small problems. It can also relieve us of the moving, blinking, wailing, shuddering, shaking, noisy urban stimuli that keep our arousal systems jacked up. Our brains get a break from registering dozens of simultaneous sensations, any of which could mean danger. Exposure to green spaces is associated with lower stress and healthier regulation of daily cortisol secretions.[10] People in England who are economically deprived have almost double (93 percent) the early mortality of the wealthiest in their country—*except* when they live in neighborhoods surrounded by greenery. Then their relative mortality dips, so that they are only 43 percent more likely to die early from any cause.[11] Nature halves their comparative risk. It's still a sad statistic about poverty, but it leads us to believe that the greenery-telomere connection deserves more exploration.

CAN MONEY BUY LONGER TELOMERES?

You don't need to be rich to have long telomeres, but having enough money for basic needs does help. One study of around two hundred African American children in New Orleans, Louisiana, found that poverty was associated with shorter telomeres.[12] Once you have basic needs met, having more money doesn't seem to help further—there are no consistent relationships between gradients of how much money you make and your telomere length. But with education, there does appear to be a dose-response gradient—the more education, the longer the telomeres.[13] Educational level is one of the most consistent predictors of early disease, so these results aren't too surprising.[14]

In a UK study, occupation mattered more than other indicators of social status: White-collar jobs (versus manual labor) were associated with longer telomere length. This was true even among twins who were raised together but as adults had different occupational status.[15]

CHEMICALS THAT ARE TOXIC TO YOUR TELOMERES

Carbon monoxide: It's odorless, flavorless, and colorless. Deep underground, in coal mines, it can build up without detection, especially after an explosion or fire. At high enough levels, it can cause a miner to asphyxiate. So in the early 1900s, miners began carrying caged canaries down into the mines with them. The miners considered them friends and would sing to the birds as they worked. If there was carbon monoxide in the mine, the canaries would show distress by swaying, reeling, or falling off their perches. The miners would know that the mine was contaminated, and they would either exit or use their breathing apparatuses.[16]

Telomeres are the canaries in our cells. Like those caged birds, telomeres are captive inside our bodies. They are vulnerable to their chemical environment, and their length is an indicator of our lifelong exposure to toxins. Chemicals are like litter in our neighborhoods—they are a part of our physical surroundings. And some are silent poisons.

Let's start with pesticides. So far, seven pesticides have been linked to significantly shorter telomeres in agricultural workers who apply them to crops: alachlor; metolachlor; trifluralin; 2,4-dichloro-phenoxyacetic acid (also known as 2,4-D); permethrin; toxaphene; and DDT.[17] In one study, the greater the cumulative exposure to the pesticides, the shorter the telomeres. It wasn't possible to determine whether one type of pesticide alone was worse or better for telomeres than any others; the study looked at an aggregate of all seven. Pesticides cause oxidative stress—and oxidative stress, when it accumulates, shortens telomeres. This study is supported by another finding, in which agricultural workers who are exposed to a mixture of pesticides while working in tobacco fields have been found to have shorter telomeres.[18]

Fortunately some of these chemicals have been banned in parts of the world. For example, there is a worldwide ban on the agricultural use of DDT (although it is still used in India). Once released, however, these chemicals don't just disappear. They live on and on in the food chain ("bioaccumulation"), so any hope to live completely free of chemicals is impossible. There are probably many toxic chemicals in small amounts in each of our cells. They end up in breast milk as well, although the benefits of breast-feeding are thought to far outweigh the exposure to chemicals. Unfortunately, many compounds on the toxic list (alachlor; metolachlor; 2,4-D; permethrin) are still used in farming and gardening and are still being produced at high levels.

Another chemical, cadmium, is a heavy metal with weighty effects on our health. Cadmium is found mostly in cigarette smoke, though we all carry low but potentially toxic levels around in our bodies because of our contact with environmental contributors such as house dust, dirt, the burning of fossil fuels such as coal or oil, and the incineration of municipal waste. Cigarette smoking has been linked to shorter telomeres—no surprise there, given the other dangerous effects of smoking.[19] Some of that relationship is due to cadmium.[20] Smokers have twice the levels of cadmium in their blood compared to nonsmokers.[21] In some countries and industries, people are exposed to cadmium through factory work. In an electronic waste-recycling town in China with known high cadmium pollution, higher cadmium levels in blood were linked to shorter telomeres in placentas.[22] In a large study of U.S. adults, those with the worst cadmium exposure have up to eleven additional years of cellular aging.[23]

Lead is another heavy metal to watch out for. Lead is found in some factories, some older homes, and developing nations that do not yet regulate lead paint and still use leaded gasoline, and it is another potential culprit in telomere shortening. Although the study of the electronic waste-recycling plant found no association between lead levels and telomere length, another study of Chinese battery factory workers who were exposed to lead as part of their work environment found some striking relationships.[24] In this study of 144 workers, almost 60 percent had lead levels high enough to meet the definition of chronic lead poisoning, and they had significantly shorter immune cell telomere length than those with normal or lower lead levels. The only difference between groups was that the group with poisoning had worked at the factory longer. Fortunately, once the lead poisoning was discovered, victims were hospitalized and given treatment (lead chelation therapy). During treatment, urine was assessed for how much lead was excreted, a measure called the "total body burden" of lead. Body burden indicates long-term lead exposure. The greater the body burden of lead, the shorter the telomeres. The correlation was .70, which is very high (the highest a correlation can be is 1). This relationship was so strong that the usual relationships of telomere length with age, sex, smoking, and obesity were not detectable in those exposed to lead. Lead exposure overrode all these factors.[25]

While severe occupational hazards have the strongest effects, it is alarming that households can also carry genotoxic hazards. Older homes may still have lead paint, which can be a danger if the paint is peeling. Many cities still use lead pipes, and the lead can travel into our homes and drinking water. Consider the tragic and shameful crisis in Flint, Michigan, in which the water supply is so corrosive that lead was leached from the pipes. The water became highly contaminated—and so did the residents' blood. While this disturbing drama has unfolded publicly on our screens, the same problem is silently taking place in many other cities that use old pipes. Particularly troublesome is that children are more sensitive to lead than adults. In one study, eight-year-old children with lead exposure had telomeres that were shorter than those of children without lead exposure.[26]

One category of chemicals, **polycyclic aromatic hydrocarbons (PAHs)**, is airborne, which makes it especially hard to avoid. PAHs are by-products of combustion and can be breathed in from fumes from cigarette and tobacco smoke, coal and coal tar, gas stoves, wildfires, hazardous waste, asphalt, and traffic pollution. You can also be exposed to PAHs if you eat foods grown in affected soil or that have been cooked on a grill. Beware. Higher exposure to PAHs has been shown to be associated with shorter telomere length in several studies.[27] An investigation of PAHs offered a caution for pregnant women: the closer a pregnant mother lived to a major roadway, and the fewer trees and plants in her neighborhood (which can reduce air pollution levels), the shorter were the telomeres of her placenta, on average.[28]

CHEMICALS, CANCER, AND *LONGER* TELOMERES

Some chemicals are associated with *longer* telomeres. This may sound good, but remember that very long telomeres in some cases are associated with uncontrolled cell growth—in other words, cancer. So when genotoxic chemicals get into our bodies, we are more likely to get mutations and cancerous cells, and if the telomeres of those cells are long, they are more likely to divide and divide and divide into cancerous tumors. This is one reason we are so concerned about the widespread use and marketing of supplements and other products that claim to lengthen your telomeres.

We are concerned that chemical exposures and telomerase activating supplements may damage cells, or that they may increase telomerase and change telomeres in radical or inappropriate ways that our bodies have not learned to cope with. But when you practice naturally healthy habits such as stress management, exercise, good nutrition, and good sleep, your telomerase efficiency increases slowly, steadily, and over time. This natural process protects and maintains your telomeres. In some cases, lifestyle changes may even help your telomeres grow a bit longer, but in a way that won't trigger uncontrolled cell growth. Healthy lifestyle factors that have been correlated with longer telomeres have *never* been shown to increase cancer risk. Lifestyle changes influence telomeres through mechanisms that are different from and safer than chemical exposures or supplements.

Which chemicals might unnaturally lengthen telomeres too much? Exposure to **dioxins and furans** (toxic by-products that are released through various industrial processes and that are commonly found in animal products), **arsenic** (common in drinking water and some foods), **airborne particulate matter**, **benzene** (exposure occurs via tobacco smoke as well as gasoline and other petroleum products), and **polychlorinated biphenyls** (or PCBs, a class of banned compounds that is still found in some high-fat animal products) is associated with longer telomere length.[29] What is so interesting is that some of these chemicals have also been linked to cancer risks. Some have been linked to higher rates of cancer in animals; others have been studied in labs, where heavy doses are put into cells and create cancer-promoting molecular changes. It is possible that chemicals can both create fertile ground for mutations and cancerous cells, and create high telomerase or longer telomere length, promoting greater likelihood the cancerous cells might be replicated. We speculate that telomeres thus may be one link in the chemical–cancer relationship.

To put this into perspective, the American Association for Cancer Research Cancer Progress Report of 2014 informs us that 33 percent of the relative contribution to overall risks of developing cancer is from tobacco use alone, and about 10 percent is attributable to occupational and environmental exposures to pollutants.[30] But that low percentage is for the United States; it is not known how much higher it is in countries and regions of the world where environmental pollution and exposures at work are much less well controlled. Furthermore, a 10 percent increase in risk may seem small, but since there are over 1.6 million new cases of cancers every year in the United States alone, that 10 percent translates into 160,000 new

Telomere Toxins

Chemicals Linked to Shorter Telomeres	Chemicals Linked to Longer Telomeres (Long telomeres in these conditions indicate a possible risk of uncontrolled cell growth and some forms of cancer.)
Heavy metals, such as cadmium and lead	Dioxins and furans Arsenic Particulate matter Benzene PCBs
Agricultural pesticides and lawn products: alachlor metolachlor trifluralin 2,4-dichlorophenoxyacetic acid (also known as 2,4-D) permethrin Mostly no longer produced but still present in the environment: toxaphene DDT	
Polycyclic aromatic hydrocarbons (PAHs)	

cancer cases per year. Think about it. Every year, 160,000 additional people and their families have their lives irrevocably changed by a diagnosis of cancer. And that's just in the United States; the World Health Organization estimates that there are 14.2 million new cases of cancer around the globe each year, so we could estimate that 1.4 million new cases of cancer each year come from environmental pollution.[31]

PROTECT YOURSELF

What can you do? More research is needed to fully understand the connection between these chemicals and cell damage, but in the meantime it is reasonable to take all the precautions you can. I have always had a preference to use natural products—but only when it was convenient for me to buy them. After realizing that so many of our household cleaners and cosmetics contained genotoxic and telomere-damaging chemicals, I now actively seek out natural products.

You may also want to change the way you eat and drink. Arsenic is naturally found in wells and groundwater, so you can either have your water tested or use a filter. Avoid plastic drinking bottles and cookware. Even BPA (Bisphenol A)–free plastic bottles may not be free of other harmful chemicals. BPA substitutes may be as unsafe; they just haven't been studied to the same extent (plus, we may soon have more plastic in the ocean than fish if we don't reduce our reliance on plastic bottles). Try not to microwave plastics, even the ones that say they are microwavable. It's true that microwaveable plastics won't warp when you heat them, but there are no promises that you won't get a dose of plastics in your food.

How can you reduce your exposure to smoke, air, and traffic pollution? Avoid living near major roadways if possible. Don't smoke (yet another good reason to quit), and avoid passive smoke. Greenery—trees, green space, and even house plants—can help reduce the levels of air pollutants inside your home and in a city, including volatile organic compounds. There is no direct evidence that living with more plants leads to longer telomeres, but there are correlations to suggest that increasing your exposure to greenery can be protective. Try to walk in parks, plant trees, and support urban forestry.

For more ways to protect yourself, see the Renewal Lab on page 276.

FRIENDS AND LOVERS

Long ago, when most of humanity lived in tribes, each group would delegate a few of its members to keep watch at night. The folks on watch would remain alert for fires, enemies, and predatory animals, and everyone else could sleep soundly, knowing that they were protected. In those perilous days, belonging to a group was a way to ensure your safety. If you couldn't trust your night watchmen, you weren't going to get your much-needed sleep. Our ancestors' version of poor social capital and lack of trust!

Flash-forward to contemporary life. When you lie down in your bed at night, you probably don't worry very much about panthers dropping on you from above, or enemy warriors skulking behind the drapes. Nevertheless, the human brain hasn't changed much since tribal days. We're still wired to need someone around who "has our backs." Feeling connected to others is one of the most basic human needs. Social connection is still one of the most effective ways to soften the danger signal; its absence will amplify it. That's why it feels so good to belong to a cohesive group. It feels good to be in connection with others—to give or receive advice, borrow or lend something, work together, or share tears and feel understood. People with relationships that allow for this kind of mutual support tend to have better health, whereas people who are socially isolated are more stress reactive and depressed, and are more likely to die earlier.[32]

In animal research, even rats, who are social animals, suffer when they are caged solo. Little did we know how stressful isolation is for this social animal. Now we know that when rats are caged alone, they don't receive the safety signals from being in close proximity to others and feel more stressed out. They get three times more mammary tumors than the rats who live in a group.[33] The rats' telomeres weren't measured. But a similar experiment found that parrots caged alone have faster telomere shortening than when they are with a mate.[34]

Aside from my bicycling disappointment, I (Liz) was generally happy as a postdoctoral fellow at Yale. But as it became time to think about finding a job, I began to worry. I'd wake up in the night in a cold sweat of anxiety, wondering how on earth I would ever become employed. One of the hurdles I had to overcome was preparing a job seminar, a lecture that I'd deliver when interviewing for academic positions. Feeling insecure, I overdid it. Desperate to convince a skeptical world about the validity of my scientific conclusions, I poured every bit of my data into the text. When I practiced the talk in front of my colleagues, the reaction was . . . muted, to say the least. The lecture was so dense that it was unintelligible. I went back to my shared office and succumbed to despairing tears. The head of the lab, Joe Gall, came by and offered kind, encouraging words. That helped. Then Diane Juricek (later Lavett), dropped in. Diane was a visiting junior professor working in a neighboring lab, and she and I shared group meetings and lunch tables. Diane volunteered to help me work my talk into shape, taking out the excessive quantities of data description and forming it into a more coherent whole. Then she helped me rehearse the lecture in the big, old-fashioned hall near the building where we worked. This enormous generosity to a younger, less experienced colleague—Diane didn't even know me well—made a huge impression on me. I realized what an academic scientific community could be about.

At the time, I was simply grateful for Diane's help. I didn't know then that my cells were likely responding to the support. Good friends are like the trusted night watchmen; when they're around, your telomeres are more protected.[35] Your cells beam out fewer C-reactive proteins (CRPs), proinflammatory signals that are considered a risk factor for heart disease when they appear in high levels. [36]

Do you have someone in your life who is close to you but also causes unease? About half of all relationships have positive qualities with less helpful interactions, what researcher Bert Uchino calls "mixed relationships." Unfortunately, having more of these mixed-quality relationships is related to shorter telomeres.[37] (Women with mixed friendships have telomeres that are shorter; both women and men have shorter telomeres when the mixed relationship is with a parent.) That makes sense. These mixed relationships are characterized by friends who don't always know how to offer support. It's stressful when a friend misunderstands your problems or doesn't give you the kind of support you really want. (For example, a friend may decide you need a long pep talk when what you really need is a shoulder to cry on.)

Marriages come in all flavors, and the better the quality of the marriage the better the health benefits, although these are what we consider statistically small effects.[38] Put someone from a satisfying marriage into a difficult situation, and they'll likely have more resilient patterns of stress reactivity.[39] Happily married people also have a lower risk of early mortality. Marriage quality has not been examined with telomere length yet, but we do know that married people, or people living with a partner, have longer telomeres.[40] (This was a surprise finding from a genetic study of 20,000 people, and the relationship was stronger in the older couples.)[41]

Sexual intimacy in marriage may matter for telomeres, too. In one of our recent studies, we asked married couples if they had been physically intimate during the previous week. Those who answered yes tended to have longer telomeres. This finding applied to both women and men. This effect could not be explained away by the quality of the relationship or other factors relating to health. Sexual activity declines less in older couples than stereotypes would have us believe. Around half of married thirty- to forty-year-olds, and 35 percent of sixty- to seventy-year-olds, engage in sexual activity anywhere between weekly and a few times a month. Many couples remain sexually active well into their eighties.[42]

Couples in unhappy relationships, on the other hand, suffer from a high level of "permeability"—they pick up on each other's stress and negative moods. If one spouse's cortisol rises during a fight, so does the cortisol of the other spouse.[43] If one spouse wakes up in the morning with a big stress response, the other is more likely to as well.[44] Both are operating at a high level of distress, leaving no one in the relationship who can put the brakes on the tension, no one who can say, "Whoa, wait. I see you're upset. Let's take a breath here and talk about it, before things get out of control." It's easy to imagine that these relationships are wearing and depleting. Our physiological responses moment to moment are more synced with our partner's than we may realize. For example, in one study examining couples having both positive and stressful discussions in the lab, heart rate variability followed the pattern of the other partner with a slight lag.[45] We suspect the next generation of research on relationships is going to reveal many more ways that we are connected physiologically to people we are close to.

RACIAL DISCRIMINATION AND TELOMERES

One Sunday morning, thirteen-year-old Richard decided to attend a friend's church in a town a few miles outside his Midwestern city. "I guess there weren't too many black people at the church to begin with," says Richard, who is black, "and I guess the two of us were dressed differently." Richard sat quietly with his friend in the reception area, waiting for the service to begin. As a minister's son, Richard had grown up in churches; he had always known them to be places where he felt welcomed, accepted, and safe. Then a woman who ran one of the church programs walked up to them.

"What are you guys doing here?" she asked in a pointed tone. They explained that they were planning to attend the Sunday service.

"I don't think you're in the right place," she said, and told them to leave.

"I felt so uncomfortable," Richard recalls now of the incident. "She kind of convinced me that I actually didn't belong there. We ended up leaving the church and not going to the service. I almost couldn't believe it had happened, but then my dad e-mailed the minister, and he confirmed that the details were correct. The woman really had said all those things. It seems inhuman that people would go to such lengths to get me out of a church."

Discrimination is a serious form of social stress. Discriminatory acts of any kind, whether they target sexual orientation, gender, ethnicity, race, or age, are toxic. Here we're zeroing in on race, because that is where telomere research has been focused. In the United States, being black, and especially being a black man, means you are more vulnerable to encounters like the one Richard had. He says, "When I talk about racism, people think I mean something extreme. But it can be small, like when a white mother grabs her child's hand when an African American teenager walks by. It hurts."

Unfortunately, racism in its extreme form is also common. African American men are more likely to be accused of a crime and attacked by the police. Now, given dashboard cameras and iPhones, we see these painful scenes on our TVs often. Police officers are like every other human: they make automatic judgments about people from a visibly different social group. Meet someone new, and within milliseconds your brain is assessing whether the person is "same" or "other." Does the person look like me? Is he or she familiar in some way? When the answers are yes, we instinctively judge the person as being warmer, more friendly, and more trustworthy. When the person seems different from us, our brains judge them to be potentially hostile and dangerous.[46]

As we said, this is an instant, unconscious reaction. It is a reason that skin color can set off automatic judgments—but it's not an excuse for acting on those judgments. All of us have to consciously work to counter this internal bias. Tim Parrish, who was raised in a close-knit but racist community in Louisiana during the 1960s and '70s, is now an adult in his fifties. Tim, who is white, admits that sometimes racist assumptions pop into his head, even though he doesn't want them there and no longer believes them to be true. But, as Parrish explained in a opinion piece for the New York *Daily News*, "What gets injected into us as beliefs is not fully our choice. What is our choice is to be constantly vigilant, to deconstruct the assumptions we make, to combat impulses we may have that lead us in the direction of thinking we are somehow the generalized victim and the more civilized color."[47] In a relatively low-stress situation, this mental work against bias may be easier to accomplish than in fast-moving, tense situations. It is a reason that "driving while black" means you are more likely to be pulled over. If you're a black man in America and your behavior seems dangerous, or is hard to interpret, you are more likely to be shot. My (Elissa's) husband, Jack Glaser, a public policy professor at the University of California, Berkeley, works on training police officers to reduce racial bias. He is helping to adapt police procedures so that they are not so heavily influenced by automatic judgments that can lead to racial discrimination. Although he and his academic colleagues categorize this as policy work, I think of it as stress reduction at a societal level, and possibly telomere relevant!

The amount of suffering people experience when they are targets of discrimination runs very deep. African Americans tend to develop more chronic diseases of aging. For example, they have higher rates of stroke than other racial and ethnic populations in the United States. Poor health behaviors, poverty, and lack of access to good medical care may explain some of these statistics, but so does a lifetime of greater stress exposure. In a study of older adults, African Americans who experience more daily discrimination had shorter telomeres, and this relationship did not hold up for whites (who experience less discrimination in the first place).[48] But this is probably not a simple, straightforward relationship; it may depend on attitudes we are not even aware of within ourselves.

David Chae at the University of Maryland performed a fascinating study that looked at low-income, young black men living in San Francisco. He wanted to know what happens to telomeres when people internalize the common societal bias, meaning that they come to believe society's negative opinions of them at an unconscious level. Discrimination alone had a weak effect. The men who had been discriminated against *and* internalized the disparaging cultural attitudes toward blacks had shorter telomeres.[49] Internalized bias toward blacks is tested by a computer task using reaction times to see how quickly people pair the word *black* with negative words. You can test your own bias at this website: https://implicit.harvard.edu/implicit/user /agg/blindspot/indexrk.htm. Just don't berate yourself for having automatic biases; most of us do. We suspect we will see more data on discrimination and telomeres in the coming years.

Knowing how places and faces affect your telomere health can be reassuring, or it can be unsettling. It all depends on your situation—where you live, the quality of your relationships, and how much you've internalized discrimination (discrimination toward any aspect of yourself—race, sex, sexual orientation, age, disability). But *all* of us can take steps to reduce toxic exposures, improve the health of our neighborhoods, become more aware of our biases toward other groups, and create positive social connections. The Renewal Lab at the end of this chapter offers some ways to get started.

TELOMERE TIPS

- We are interconnected in ways we cannot see, and telomeres reveal these relationships.
- We are affected by the toxic stress of discrimination.
- We are affected by toxic chemicals.
- We are affected in more subtle ways, by how we feel in our neighborhood, by the abundance of green plants and trees nearby, and by the emotional and physiological states of those around us.
- When we know how we are affected by our surroundings, we can begin to create healthful, supportive environments in our homes and our neighborhoods.

RENEWAL LAB

REDUCE YOUR TOXIC EXPOSURES

We've already described some basic precautions against plastics and pollution that could shorten—or dangerously lengthen—your telomeres. Here are some more advanced moves:

- **Eat less animal and dairy fat.** The fatty parts of meat are where certain bioaccumulative compounds collect and concentrate. The same goes for the fat in large, long-lived fish, except that there is a balancing issue to weigh. Fatty fish such as salmon and tuna also contain omega-3s, which are good for your telomeres, so eat in moderation.
- **Think about the air when you turn up the heat with meat.** If you cook meat on a grill or on a gas stove, use ventilation. Try to avoid exposing the food directly to open flames, and try not to eat the charred portions, no matter how tasty they are. A good idea for any food.
- **Avoid pesticides in your produce.** Eat foods that are free of pesticides when possible; at the very least, wash your produce thoroughly before consuming. Purchase organic fruits, vegetables, and meats, or grow your own. Consider growing lettuce, basil, herbs, and tomatoes in pots on your balcony. Safe alternatives for dealing with pesky critters can be found here: http://www.pesticide.org/pests_and_alternatives.

- **Use housecleaning products containing natural ingredients.** You can make many of these products yourself. We like the housecleaning "recipes" found at http://chemical -free-living.com/chemical-free-cleaning.html.
- **Find safe personal-care products.** Carefully read the labels on personal-care products such as soap, shampoo, and makeup. You can also visit http://www.ewg.org/skindeep to identify which chemicals are in your beauty products. When in doubt, buy products that are organic or all natural.
- **Buy nontoxic house paints.** Avoid paints that contain cadmium, lead, or benzene.
- **Go greener.** Buy more house plants: two per one hundred square feet is ideal for keeping your air filtered. Good choices include philodendrons, Boston ferns, peace lilies, and English ivy.
- **Support urban forestry** with your money or your labor. Green spaces offer so many benefits to mind and body, as well as to healthy communities. **One newer idea can be considered in dense urban megacities, where one cannot plant enough trees to rid the air of toxins.** If you live in a city, consider lobbying your municipal government to install air-purifying billboards. These billboards do the work of 1,200 trees, cleaning a space of up to 100,000 cubic meters by removing pollutants such as dust particles and metals from the air.[50]
- **Stay up to date about toxic products by downloading the "Detox Me" app by Silent Spring:** http://www .silentspring.org/.

INCREASE THE HEALTH OF YOUR NEIGHBORHOOD: SMALL CHANGES ADD UP

To brighten a corner of your own neighborhood, follow the example of our San Francisco neighbors and place a few benches and tables on a bare cement sidewalk, along with a little greenery. These "park-lets" attract neighbors and promote socializing and peaceful loung-ing. Or consider one of these:

- **Add art.** A mural or even a beautiful poster can infuse a drab area with hope, truth, faith, and positivity. Residents in a Seattle neighborhood painted boarded-up shop windows with pictures of the businesses they hoped to attract: an ice-cream parlor, a dance studio, a bookshop, and so on. The paintings helped entrepreneurs see the potential of the neighborhood. They brought their small businesses to the block, revitalizing the area and bringing economic growth to the community.[51]

- **Get greener**, especially if you are a city dweller. More green space in a neighborhood is associated with lower cortisol and lower rates of depression and anxiety.[52] Turn a vacant lot into a sustainable community fruit or vegetable patch, or plant trees and flowers in a small park space. "Greening" a vacant lot has been associated with a decrease in gun violence and vandalism and an increase in the residents' general feeling of safety.[53]

- **Warm your neighborhood tone.** Social capital is an invaluable resource that predicts good health. It's defined by the level of community engagement and positive activi-ties and resources that exist in a neighborhood, and one of its most important ingredients is trust. So be the one to make the first move. Cook or bake a little extra and drop it off at your neighbors' house on a small plate. Share vegetables or flow-ers from your garden. Help out by shoveling snow, giving a ride to an elderly person, or starting a neighborhood watch. Leave a welcoming note for newcomers to the neighborhood, or plan a block party. You could also join the trend of open-ing a Little Free Library in front of your house by putting out a wooden cabinet where books are shared. (See https://Little FreeLibrary.org.)

- **Smiles matter.** Acknowledge people you pass on the street. As social animals, we are exquisitely sensitive to social cues, noticing signs of acceptance and especially rejection. Each day, we interact with strangers or acquaintances, and we can either feel separate from them—or we can connect with them in a small way that has a positive effect. Give people an "air gaze" (looking past the face, with no eye contact) and they will tend to feel more disconnected from others. Give them a smile and eye contact, and they feel more connected.[54] Plus, when people are given a smile, they are more likely to help someone else in their next moments.[55]

STRENGTHEN YOUR CLOSE RELATIONSHIPS

Then there are the people we wake up to almost every day—our family, and colleagues we work with. The quality of these relationships is important to our health. It is easy to be neutral, to take those we see all the time for granted. Investigate what it is like to really acknowledge your close ones in a significant way:

- Show gratitude and appreciation. Say, "Thanks for doing the dishes" or "Thanks for supporting me at the meeting."
- Be present. This means not looking at a screen or around the room. Give your full, sincere attention. That is a gift you can give another person, and it doesn't cost a dime.
- Hug or touch your loved ones more often. Touch releases oxytocin.

CHAPTER TWELVE

Pregnancy: Cellular Aging Begins in the Womb

When I (Liz) found out I was pregnant, I instantly felt protective toward my tiny unborn baby. On getting back the test results, I immediately stopped smoking. Luckily, I had been smoking only lightly, a few cigarettes a day at most. I found the transition easy to make—especially because I was so concerned about the baby's wellbeing. I have never smoked again. I also became very interested in what to eat. Listening to my obstetrician and her team, I paid attention to getting nutrients from foods (like fish, chicken, and leafy greens). I also took the micronutrient supplements for iron and vitamins they recommended.

Now, many years later, we have a much deeper understanding of how a mother's nutrition and health status affects her developing baby. We are also learning what happens to a baby's telomeres in the womb. Little did I suspect, all those years ago, that my decisions may have helped to protect my baby's telomeres. Or, more spectacular, that the choices I made—and the events that had happened to me years before the baby was born—might even have affected the starting point of my son's telomeres.

Telomeres continue to be shaped throughout adulthood. Our choices can make our telomeres healthier, or they can hasten their shortening. But long before we're old enough to make decisions about what to eat or how much to exercise, and before chronic stress starts to threaten our DNA base pairs, we begin life with an initial telomere setting. Some of us arrive in this world with shorter telomeres. Some of us are lucky to have longer ones.

As you can imagine, telomere length at birth is influenced by genetics, but that is not the whole story. We are learning astonishing things about how parents can shape their children's telomeres—before those children are even born. And this matters—the telomere length at birth and early childhood is a major predictor of what we have left as we become adults.[1] The nutrients that a pregnant mother consumes, and the level of stress she experiences, can influence her baby's telomere length. It is even possible that parents' life histories can affect telomere length in the next generation. In a sentence: Aging begins in utero.

PARENTS CAN PASS THEIR SHORTENED TELOMERES TO THEIR CHILDREN

Chloe, now age nineteen, became pregnant two years ago. Without much support or understanding from her parents, she left home and moved in with a friend. To help pay her share of the rent, she dropped out of high school and began working a minimum-wage retail job. Despite her difficult circumstances, Chloe has been determined to give her baby a good start in life. While she was pregnant, she did her best to get prenatal care. She took the prenatal vitamins she was prescribed, even though she says they made her sick. When her son was born, Chloe pledged that he would always, always feel loved.

Chloe is determined to give her child what she didn't have—better health and greater satisfaction—and to help lift him as part of the next generation. But there is shocking evidence that Chloe's low education level could have indirectly shaped her baby's telomeres—*while he was still in the womb*. Babies whose mothers never completed high school have shorter telomeres in their cord blood compared to those whose moms had a high school diploma—meaning that they have shorter telomeres from the first day of their lives.[2] Older children whose parents have lower levels of education also have shorter telomeres.[3] These findings are based on studies that controlled for other factors that could have influenced the results, such as, in the baby study, whether their babies had a low birth weight.

Let this sink in for a moment, because the implications, if borne out in subsequent studies, are revolutionary. How could a parent's education level affect the telomeres of her developing baby?

The answer is that telomeres are transgenerational. Parents can, of course, hand down *genes* that affect telomere length. But the really profound message is that parents have a second way of transmitting telomere length, known as *direct transmission*. Because of direct transmission, both parents' telomeres—at whatever length they are at the time of conception in the egg and sperm—are passed to the developing baby (a form of epigenetics).

Direct transmission of telomere length was discovered when researchers were investigating telomere syndromes. Telomere syndromes, as you'll remember, are genetic disorders that lead to hyperaccelerated aging. Their victims have extremely short telomeres. People with telomere syndromes—think of Robin in an earlier chapter—often watch their hair turn gray while still in their teens. Their bones can become fragile, or their lungs can stop working properly, or they can develop certain cancers. In other words, they make an early and tragically dramatic entrance into the diseasespan. Telomere syndromes are inherited, caused when parents pass a single mutated telomere-related gene down to their children.

But there was a mystery. Some children in these families are lucky enough not to inherit the bad gene that causes the telomere syndrome. You'd think these children would escape premature cellular aging, right? Yet some children without the bad gene *still* showed mild to moderate signs of early aging—not as severe as what you might find in a full-blown telomere syndrome, but beyond what is normal, such as very early graying. Researchers decided to measure these children's telomeres and found that their telomeres were, in fact, unusually short. The children had escaped the gene that causes inherited telomere syndrome, but somehow they had still been born with short telomeres that persisted in being short. These children had received short telomeres from their parents—but not through inheritance of a bad gene. Although the children were growing up with normal telomere maintenance genes, because their telomeres had started off so short, the telomeres simply could not be replenished fast enough to catch up and attain normal lengths.[4]

How can this be? How can children receive short telomeres from their parents, if not through genes? The answer, once you know it, is immediately obvious. It turns out that parents can directly transmit their telomere length to the child in the womb. Here's how it happens: A baby begins with a mother's egg, fertilized by the father's sperm. That egg contains chromosomes. Those chromosomes contain genetic material, of course. That's how genetic material is passed down to the baby. But the material of the chromosomes of the fertilized egg also includes the telomeres at their ends. Because the baby is made from the egg, the baby receives those telomeres directly—of whatever length they are at that time. **If the mother's telomeres are short throughout her body (including those in the egg) when she contributes the egg, the baby's telomeres will be short, too. They'll be short from the moment the baby starts developing.** That's how children without the bad gene received shorter telomeres. And this suggests that if the mother has been exposed to life factors that have shortened her telomeres, she can pass those shortened telomeres directly to her baby. On the other hand, a mother who has been able to keep her telomeres robust will pass her stable, healthy telomeres to the growing child.

What does Dad contribute? Upon fertilization of the egg, the chromosomes that come in from the dad via the sperm join the chromosomes from the mother. The sperm, like the egg, also bears its own telomeres that are directly transmitted to the developing baby. The research to date suggests a father *can* directly transmit short telomeres, but just not to the extent that a mother with very short telomeres would. In a new study of 490 newborns and parents, babies' cord blood telomeres were more related to their moms' telomere length than their dads', but they are both clearly influential.[5]

So far, there have been only a few studies that look at direct transmission of telomere length in humans. That would involve measuring both the genetics for telomeres, and the telomeres themselves, so we can separate out the effects of genetics from life experience. Those studies have all been focused on families with telomere syndromes.[6] But we and other researchers suspect that it happens in the normal population, too.[7] As you're about to see, the science of direct transmission suggests a way that poverty and disadvantage may have effects that echo through the generations.

CAN SOCIAL DISADVANTAGE BE PASSED DOWN THROUGH THE GENERATIONS?

Did your parents suffer from prolonged, extreme stress before you were born? Were they poor, or did they live in a dangerous neighborhood? You already know that the way that your parents lived before you were conceived probably affected their telomeres. It may have also affected *yours*. If your parents' telomeres were shortened by chronic stress, poverty, unsafe neighborhoods, chemical exposures, or other factors, they may have passed their shortened telomeres to you through direct transmission in the womb. There is even the possibility that you, in turn, could pass those shortened telomeres to your own children.

Direct transmission has strong and chilling implications for all of us who care about future generations. It raises a controversial idea. In our view, the evidence from telomere syndrome families suggests that it is possible for the effects of social disadvantage to accumulate over the generations. We can already see the pattern in large epidemiological studies: Social disadvantage is associated with poverty, worse health—and shorter telomeres. Parents whose telomeres are shortened by this disadvantage may directly transmit those shorter telomeres to their babies in utero. Those children will be born a step behind, or base pairs behind, with telomeres shortened by their parents' life circumstances. Now imagine that as these children grow up, they are also exposed to poverty and stress. Their telomeres, already shortened, will erode even further. In a downward spiral, each generation directly transmits its ever-shortening telomeres to the next. And each new baby could be born further and further behind, with cells that are more and more vulnerable to premature aging and an early diseasespan. This pattern is exactly what happens in the rare telomere syndrome families: With each successive generation, the progressively shorter and shorter telomeres cause earlier and worse disease impacts than in the generation before.

From the first moments of life, telomeres may be a measure of social and health inequalities. They may help explain the disparity among different postal codes in the United States. People living in certain ZIP codes that represent wealthier areas have life expectancies up to ten years longer than people in other ZIP codes that cover poorer areas. This difference has often been explained by risky behaviors or exposure to violence. But the actual biology of babies

Figure 26: Aging at Birth? "Mom, what happened to the level playing field?" Babies are born with short telomeres depending on their mothers' genes but also their mothers' biological health, level of stress, and, possibly, level of education.

born into these neighborhoods may also be different. Tragically, a neighborhood's health challenges may be compounded from generation to generation. But biology is not destiny; there are many things we can do to maintain our telomeres through our own lifetime.

NUTRITION IN PREGNANCY: FEEDING A BABY'S TELOMERES

"You're eating for two now." Pregnant women hear this advice all the time. It's true: A developing baby gets its calories and nutrition from the food the mother eats (and it's not true the mom needs to eat twice as much). Now it appears that what a pregnant woman eats can also affect her baby's telomeres. Here, we'll look at the nutrients that have been connected to telomere length in utero.

Protein

Animal research suggests that modest protein deprivation in pregnancy causes accelerated telomere shortening in the offspring in a number of tissues, including the reproductive tract, and can lead to earlier mortality.[8] When a mother rat is fed a low-protein diet during pregnancy, her daughters have shorter telomeres in their ovaries. They also have more oxidative stress and higher mitochondria copy numbers, suggesting that the cells are under high stress and to cope they are rapidly producing more mitochondria.[9]

The damage can even travel to the third generation. When the researchers looked at the granddaughter rats, they found that their ovaries had undergone accelerated tissue aging. They had more oxidative stress, higher mitochondria copy numbers, and shorter telomeres in their ovaries. The granddaughters were victims of early cellular aging, all as a result of a low-protein diet two generations earlier.[10]

Co-enzyme Q

There is strong evidence in humans and animal models that maternal malnourishment during pregnancy leads to increased risk of heart disease in the offspring. If a pregnant mother doesn't get enough to eat, or isn't adequately nourished, her child may be born at a low birth weight. Often, there's a rebound effect, with the underweight baby playing a game of catch-up that eventually leads to overeating and obesity. Babies born at a low birth weight carry an increased risk for cardiovascular disease as they get older, and babies who experience this postnatal rebound of rapid weight gain have a risk that's even higher.

As we said, this scenario links maternal malnourishment to heart disease—and one of the links in the chain may be telomere shortening. Rat pups that are born to mothers who don't get enough protein tend, like their human counterparts, to have a low birth weight. And just like human babies, they often experience a later rebound of weight gain. Susan Ozanne at the University of Cambridge has found that these rat pups have shorter telomeres in the cells of several organs, including the heart aorta. They also have lower levels of an enzyme known as CoQ (ubiquinone). CoQ is a natural antioxidant that is found mostly in our mitochondria, which play a role in energy production. A CoQ deficit has been associated with faster aging of the cardiovascular system. But when the pups' diets were supplemented with CoQ, the negative effects of protein deprivation were wiped out, including the effects on telomeres.[11] Ozanne and her colleagues concluded that "early intervention with CoQ in at-risk individuals may be a cost-effective and safe way of reducing the global burden of [cardiovascular diseases]."

Of course, it's a long leap from rat to human. What's good for one may not be good for another. Even in rats, we don't know whether the benefits are restricted to pups whose mothers were deprived of protein. CoQ should be put on the list of nutrients for further study of their potentially positive effects on telomeres. If those benefits exist, they could be harnessed for babies of mothers who had inadequate nutrition during pregnancy, or even for adults who are at risk for heart disease. Note that no studies we are aware of have used CoQ during pregnancy, or examined the safety, and thus we are not recommending it.

Folate

Folate, a B vitamin, is another crucial nutrient during pregnancy. You probably know that folate decreases the risk of spina bifida, a birth defect, but it also prevents DNA damage by shielding the regions of chromosomes known as the centromere (all the way in the middle of the chromosome) and the subtelomere (the chromosome region just inside and next to the telomere). When folate levels drop too low, the DNA becomes hypomethylated (losing its epigenetic marks), and the telomeres become too short—or, in a few cases, abnormally elongated.[12] Low folate levels also cause an unstable chemical, uracil, to be incorporated into the DNA, and possibly into the telomere itself, perhaps causing temporary elongation.

Babies of mothers who have inadequate folate during pregnancy have shorter telomeres, further pointing to folate as vital for optimal telomere maintenance.[13] And gene variants that make it harder for the body to use folate are associated with shorter telomeres in some studies.[14]

The U.S. Department of Health and Human Services recommends that pregnant women get between 400 and 800 micrograms of folate daily.[15] Just don't assume that getting even *more* folate is better. At least one study hints that a mom overdoing vitamin supplementation of folate may decrease her baby's telomere length.[16] To repeat a theme of this book: moderation and balance are essential!

BABY'S TELOMERES ARE LISTENING TO MOM'S STRESS

A mother's psychological stress may affect her developing baby's telomere length. Our colleagues Pathik Wadhwa and Sonja Entringer from the University of California, Irvine, asked if we would collaborate on a study of prenatal stress and telomeres, and we were delighted to join them and study the start of life. The study was small, but it showed that when mothers experience severe stress and anxiety during pregnancy, their babies tend to have shorter telomeres in their cord blood.[17] A baby's telomeres can suffer from prenatal stress. A recent study extended this finding by examining stressful life experiences. Researchers added up the stressful events that happened in the year before giving birth. The mothers with the highest number of stressful life events had babies with telomeres that were shorter by 1,760 base pairs at birth.[18]

Sonja and Pathik wanted to know how long the effect of prenatal stress on the baby might last. They recruited a group of adult men and women and asked if their mothers had experienced any extremely stressful events while they were pregnant. (The volunteers interviewed their mothers about major events, such as the death of a loved one or a divorce.) As adults, the volunteers who had been exposed to prenatal stress were different in a several ways—even after controlling for factors that might influence their current health. They had more insulin resistance. They were more likely to be overweight or obese. When they underwent a lab-stressor test, they released more cortisol. When their immune cells were stimulated, they responded with higher levels of proinflammatory cytokines.[19] Finally, they had shorter telomeres.[20] A pregnant mother's severe psychological stress appears to have echoes into the next generation, affecting the trajectory of telomere length for decades of the child's life.

We're speaking of very serious stress here. Almost all pregnant mothers experience some mild to moderate stress—not necessarily because they are pregnant but because they are human. At this point, there is no reason to believe that these lower levels of stress are harmful to a baby's telomeres.

The main player that has been examined in pregnancy stress is cortisol. This hormone is released from the mother's adrenal glands and can cross the placenta to affect the fetus.[21] In birds, cortisol from a stressed pregnant bird will make its way into the egg to affect the offspring. Either injecting cortisol into the egg or stressing out the mother can lead to shorter telomeres in chicks. These studies suggest

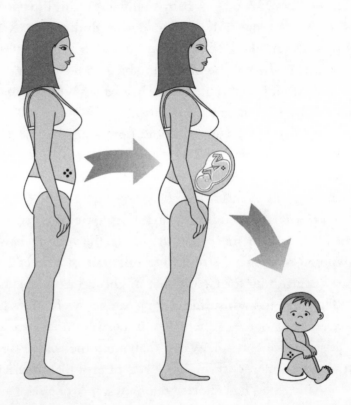

Figure 27: Telomere Transmission. There are at least three paths for telomere transmission from a parent to a grandchild. If a mother has short telomeres in her eggs, those short telomeres can be transmitted directly to the baby (this is known as germline transmission). All the baby's telomeres would then be shorter, including his or her own germline cells (sperm or eggs). During fetal development, maternal stress or poor health can lead to telomere loss in the baby, thanks to excessive cortisol exposure and other biochemical factors. Postnatally, the child's life experiences can shorten his or her telomeres. This child's short telomeres in germline can then be transmitted to his or her future offspring. Mark Haussman and Britt Heidinger have described such transmission pathways in animals and humans.[22]

the possibility that a human mother's stress could be passed on to her baby in the form of short telomeres. Again, what can happen in birds may not happen in humans—but we know enough about chronic stress and telomeres to state that pregnant women must be protected from life's harshest stressors. These include any kind of emotional or physical abuse, violence, war, chemical exposures, food insecurity, and grinding poverty. At the very least, we can support local efforts to provide services and support that buffer pregnant women from survival threats like hunger and violence from the earliest days of pregnancy.

It's clear that parents, especially mothers, influence the telomere health of their babies. And as you're about to see, telomere health is also heavily determined by the way we raise our children and teenagers.

While the health of future generations is important to all societies, it is not, in reality, often paid attention to. Our investment in our most vulnerable young citizens can now be thought of in terms of also investing in base pairs of telomeres, for our collective future of robust health and extended healthspans.

TELOMERE TIPS

- Some transmission of telomere length is out of our control. This includes genetics and direct transmission from eggs and sperm. Telomere transmission to children can happen when a parent has very short telomeres, regardless of genetics. It is a real possibility that we could be unknowingly transmitting disparities in health through this direct telomere transmission.
- Some of what we transmit is under our control. A mother's severe stress during pregnancy, smoking, and intake of certain nutrients, such as folate, are related to her baby's telomere length.
- The transmission of severe social disadvantage through telomeres can likely be blocked through policies that protect the health of women of childbearing age, and especially pregnant women, from toxic stressors and food insecurity.

291

RENEWAL LAB

GREENING OF THE WOMB

Pediatrician Julia Getzelman of San Francisco recommends that pregnant mothers think about "greening the womb" as well as their house. If you are pregnant, review our Renewal Lab ideas for minimizing chemical exposure in the previous chapter (page 276). Here are some key ways to think of greening the womb:

- Avoid **negative stress**, such as toxic relationships in which you know there will be conflict, unrealistic deadlines, and other situations in which you will not get enough sleep or be able to eat well for days. Life happens, including major events, while you are pregnant, but try to control what you can, and prioritize supportive relationships.
- Increase **well-being time**. Take prenatal yoga classes, or use a yoga video. Find ways to socialize with other pregnant women. Enjoy going out for walks, preferably in green areas.
- Eat a "rainbow" by consuming foods in a variety of rich, deep colors. Amp up **protective nutrients for a healthy developing baby**: Ensure adequate dietary protein, vitamin D_3 and B vitamins, including folate and B_{12}, fish or a high quality omega-3 fatty acid supplement, and probiotics.
- **Avoid pesticides and chemicals in food by eating an organic diet.** Limit your consumption of large and farmed

fish, which often contains accumulations of heavy metals and other industrial chemicals. Limit saccharin or other artificial sweeteners, as these can cross the placenta. (The newer artificial sweeteners may do the same; we expect more and more alarming findings.) Canned foods contain BPA (bisphenol A), a significant endocrine disruptor. Stick to what nature provides and consume a whole-foods diet. Avoid packaged foods with their many questionable additives.

- **Avoid chemical exposures at home** by wet-mopping frequently, using a vinegar-and-water mixture to clean most surfaces, and checking out safer cleaning products and cosmetics here: http://www.ewg.org/consumer-guides. Additionally, plastic PVC shower curtains, perfumes, and other fragrance-containing items such as scented candles may be a significant source of toxins.

CHAPTER THIRTEEN

Childhood Matters for Life: How the Early Years Shape Telomeres

Childhood exposures to stress, violence, and poor nutrition affect telomeres. But there are factors that appear to protect vulnerable children from damage—including sensitive parenting and mild "good stress."

In the year 2000, Harvard psychologist and neuroscientist Charles Nelson walked into one of Romania's notorious orphanages, a legacy of the brutal policies of the Nicolae Ceauşescu regime. The institution housed about four hundred children, all segregated by age as well as by disability. There was a ward full of children with untreated hydrocephalus, a disorder in which the skull expands to accommodate excess fluid, and spina bifida, a defect of the spinal cord and the bones along the spine. There was an infectious disease ward that housed children with HIV and children with syphilis so advanced that it had gone to the brain. On this same day, Nelson entered a ward full of supposedly healthy children who were around two or three years old. One of these children—they'd all been given similar haircuts and clothing, so it was hard to identify them by gender—stood in the middle of the floor, pants soaking wet, sobbing. Nelson asked one of the caregivers why the child was crying.

"His mother abandoned him here this morning," she said. "He's been crying all day."

With so many children under their care, the staff had no time for comforting or soothing. Leaving newly abandoned children alone was a way for the staff to quickly extinguish unwanted behaviors like crying. Babies and toddlers were left in their cribs for days at a time, with nothing to do but stare up at the ceiling. When a stranger walked by, the children would reach their arms out through the crib railings, begging to be held. Although the children were adequately fed and sheltered, they received almost no affection, no stimulation. As Nelson and his team built a lab inside the orphanage to study the effects of early childhood neglect on the developing brain, they had to establish a behavior rule of their own to avoid adding to the residents' distress: No crying in front of the children.

What Nelson and his colleague, Dr. Stacy Drury, learned from studies at the orphanage is both heartbreaking *and* hopeful. Early childhood neglect shortens telomeres—but there are interventions that can help neglected or traumatized children, if we can catch them at a young age. Although the conditions of the orphanages in Romania have improved generally, there are still around seventy thousand orphans and fewer international adoptions to rescue them.[1] Institutional care of children is an ongoing global crisis. War, along with diseases like HIV and Ebola, rob children of their parents and have left an estimated eight million children currently housed in orphanages around the world. We can't afford to turn away from this story.[2]

It is also a story that may have relevance inside our own homes. Telomere knowledge can guide our actions as parents, illuminating a path to raising our children in a way that is healthy for their telomeres. For adults who experienced trauma as children, understanding the long-lasting cellular effects of the past can offer motivation for treating telomeres with tender care now, in the present.

TELOMERES TRACK CHILDHOOD SCARS

When you were growing up, did you have a parent who drank too much? Was anyone in your family depressed? Were you often afraid that your parents would humiliate you or even hurt you?

In a study that painted a disturbing portrait of childhood in the United States, seventeen thousand people were asked to answer a list of ten questions much like the ones above. Around half the sample had experienced at least one such adverse event or situation in childhood, and 25 percent had experienced two or more. Six percent experienced at least *four*. Substance abuse in the family was most common, then sexual abuse and mental illness. Adverse childhood events happen across all levels of incomes and education. Worse, the more events that a person ticked off on the list, particularly if the person had four or more, the more likely the person was to have health problems in adulthood: obesity, asthma, heart disease, depression, and others.[3] Those with four or more adverse events were twelve times more likely to have attempted suicide.

Biological embedding is the name for the effects of childhood adversity that lodge themselves in the body. When telomeres are measured in healthy adults who were exposed to adverse childhood events, a dose–response relationship is often seen. The more traumatic events that a person experienced back then, the shorter their telomeres as an adult.[4] Shorter telomeres are one way that early adversity embeds itself in your cells.

Those short telomeres could have searing effects on a child. If you take a group of young children with shorter telomeres and peer inside their cardiovascular systems a few years later, you'll find that they are more likely to have greater thickening of the walls of their arteries. These are *kids* we're talking about here—and for them, short telomeres could mean a higher risk of early cardiovascular disease.[5]

That damage may begin at a very young age, though it can be halted or possibly reversed if children are rescued from adversity early enough. Charles Nelson and his team compared the children living in Romanian orphanages to ones who'd left the orphanages for quality care in foster homes. The more time the children had spent in the orphanage, the shorter their telomeres.[6] Many of the orphans showed low levels of brain activity during EEG scans. "Instead of a hundred-watt light bulb," Nelson has said, "it was a forty-watt light bulb."[7] Their brains were measurably smaller, and their average IQ was 74, which put them on the borderline of mental retardation. For most of the institutionalized children, their language was delayed and in some cases disordered. Their growth was stunted; they had smaller heads; and they had abnormal attachment behavior, which affects the ability to form lasting relationships. But, says Nelson, "the kids in foster care were showing dramatic recoveries." The children who'd been moved to foster care showed remarkable gains although they had not completely caught up to the children who had never been in orphanages at all; for example, although their IQ was still below that of the never-institutionalized children in the study, it was ten or more points higher than children in the institution.[8] There seemed to be a critical period of brain development: "The kids placed in foster care before age two had improvements in many domains that were better than kids placed after age two," Nelson says.[9] Drury, Nelson, and their team have continued to track these children over the years—and even now, the adolescents who lived at the orphanage as children experience telomere shortening at an accelerated rate.

What about the telomeres of children who are exposed to conditions that, while violent, are not quite so brutal? Scientists Idan Shalev, Avshalom Caspi, and Terri Moffitt, of Duke University, took cheek swabs from five-year-old British children. (Telomeres can be obtained from buccal cells, which live in cheeks.) Five years later, when the children were ten, they swabbed the children's cheeks again. During the five years, the researchers asked the children's mothers about whether their children had been bullied, hurt by someone in their household, or witnessed domestic violence between the parents. The children who had been exposed to the most violence had the greatest telomere shortening over the five years.[10] Maybe this effect on children is short-lasting, or it can change if their life circumstances improve. We hope so. But studies of adults in which people are asked to recall whether they had early adversity also show that those who did have early adversity have shorter telomeres, revealing what may be a lifelong imprint of childhood adversity inside them.[11] In a large study of adults in the Netherlands, reporting several traumatic events as a child was one of the few predictors of having a greater rate of shortening as an adult.[12] In addition, childhood trauma, particularly maltreatment, has been related to greater inflammation and a smaller prefrontal cortex.[13]

That imprint of early trauma can change the way you think, feel, and act. People who have faced early adversity aren't as flexible in their responses to life's varied experiences. They have a higher number of bad days, and their bad days feel more stressful to them. When something good happens, they also feel more joyful.[14] This pattern isn't unhealthy in itself. It just leads to a more intense and dynamic emotional experience. However, that intensity makes it harder to ride out the transitions between emotions. People with a traumatic childhood background tend to have more difficulties in relationships. They're more likely to engage in emotional eating and addictive behaviors.[15] They're not as good at taking care of themselves. These psychological reverberations of abuse may continue to shape mental and physical health all through life. In this way, early adversity may plant the seeds for a greater rate of telomere shortening, unless these resulting patterns of behavior are halted.

ADD UP YOUR ACES (ADVERSE CHILDHOOD EXPERIENCES)

Here's a version of the ACES test, used to measure the number of adverse childhood experiences. Take it now to evaluate your own adversity in childhood.[16]

When you were a child (up to eighteen years old):

1. Did a parent or other adult in the household often or very often swear at you, insult you, put you down, or humiliate you? Or act in a way that made you afraid that you might be physically hurt?

   No _____ If Yes, enter 1 _____

2. Did a parent or other adult in the household often or very often push, grab, or slap you or throw something at you? Or ever hit you so hard that you had marks or were injured?

   No _____ If Yes, enter 1 _____

3. Did an adult or a person at least five years older than you ever touch or fondle you or have you touch his or her body in a sexual way? Or attempt or actually have oral, anal, or vaginal intercourse with you?

   No _____ If Yes, enter 1 _____

4. Did you often or very often feel that no one in your family loved you or thought you were important or special? Or that your family didn't look out for each other, feel close to each other, or support each other?

   No _____ If Yes, enter 1 _____

5. Did you often or very often feel that you didn't have enough to eat, had to wear dirty clothes, and had no one to protect you?

Or that your parents were too drunk or high to take care of you or take you to the doctor if you needed it?

No _____ If Yes, enter 1 _____

6. Was a biological parent ever lost to you through divorce, abandonment, or other reasons?

No _____ If Yes, enter 1 _____

7. Was your mother or stepmother often or very often pushed, grabbed, or slapped? Or sometimes, often, or very often kicked, bitten, hit with a fist, hit with something hard, or made the target of a thrown object? Or ever repeatedly hit for at least a few minutes or threatened with a gun or knife?

No _____ If Yes, enter 1 _____

8. Did you live with anyone who was a problem drinker or an alcoholic, or who used street drugs?

No _____ If Yes, enter 1 _____

9. Was a household member depressed or mentally ill, or did a household member attempt suicide?

No _____ If Yes, enter 1 _____

10. Did a household member go to prison?

No _____ If Yes, enter 1 _____

Total score _____

Typically, having one adverse event is not related to health, whereas having three or four events may be. If you've had several adverse childhood events, and you feel lasting imprints on your current "mindstyle" or lifestyle, don't panic. Your childhood does not have to determine your future. If for example, you developed emotional eating as a coping strategy, you can shed that as an adult. It involves understanding why that pattern developed, and that it doesn't have to be your coping solution going forward. But before you can shed the behavior, it's important to discover alternative coping that works for you, and practice healthier ways to tolerate painful feelings over and over. There are so many ways to buffer residual effects of childhood trauma. If you are still bothered by thoughts about a difficult past, it may warrant seeking help from a professional in mental health. Remember: You are not powerless, and you not alone. Caring professionals can help you undo some of the damage that you were once powerless to stop. And remember there are positive attributes still with you. For example, severe adversity is related to feeling more compassion and empathy for others.[17]

DON'T STEP ON MY PAW! THE EFFECTS OF MONSTROUS MOTHERING

Dr. Frankenstein, step aside. Today's researchers know how to take perfectly nice rats and turn them into maternal monsters. In the lab, they can "build" a rat mother who mistreats her own pups. This is a hard subject for animal lovers to process, but it's helpful reading for anyone who wants to understand the physiology of childhood adversity.

One of the more stressful circumstances for a lactating mama rat is a lack of adequate bedding. Rats don't need luxurious mattresses to be comfortable, but mother lab rats do rely on things like facial tissues and strips of paper to build a little nest for their families. Another cause of high stress for rats is moving to a new place without enough time to habituate to it. By depriving mother rats of material for bedding and by moving them abruptly to a new cage, scientists can create highly stressed animals. Think of how stressful it would be to come home from the hospital with a newborn baby and *then* be greeted by a landlord who says, "Good, you're finally here! Before you put the baby down, let me explain that we've moved you to a new house. Also, we took all your clothing and furniture to the dump. Bye!" You'll have an inkling of what the mother rats were feeling.

These stressed mother rats mistreated their pups. They dropped them. They stepped on them. They spent less time nursing, licking, and grooming—supportive maternal activities that calm rat pups and lead to long-term changes in calming their neural stress responses. The poor pups cried out loudly, signaling their distress. This abusive early environment shaded the contours of the pups' neural development. Compared to rats who were raised by nurturing mothers, these pups had longer telomeres in a part of their brains known as the amygdala, which governs the alarm response.[18] The alarm response had apparently been switched on so often that the telomeres there were strong and robust. Not exactly a sign of a happy upbringing.

Having a strong connection between the amygdala and the prefrontal cortex, which can dampen that response, is critical for good emotion regulation. Sadly, the mistreated rat pups had shorter telomeres in a part of the prefrontal cortex. We already know that severe stress causes the nerve cells of the amygdala to branch out, to enlarge and connect to the nerve cells in other parts of the brain. The opposite tends to happen in the neurons of the prefrontal cortex, so that the connection between the two areas becomes weaker, and the rats can't turn off the stress response as easily.[19]

LACK OF MOTHERING

Parental neglect is another condition that can harm telomeres. Steve Suomi of the National Institutes of Health in Bethesda, Maryland, has been studying parenting in rhesus monkeys for the past forty years. He has found that when they are raised in a nursery from birth, without their mother but socializing with peers, they show a range of problems. They are less playful, and more impulsive, aggressive, and stress reactive (and have lower levels of serotonin in their brains).[20] He wanted to examine whether they have greater telomere attrition as well. He and his colleagues recently had the opportunity to study this in a small group of monkeys. They randomized some to be raised by their mothers and the others to be raised in a nursery for the first seven months of life. When their telomeres were measured four years later, the monkeys raised by their mothers had dramatically longer telomeres, around 2,000 base pairs longer, than the nursery-raised monkeys.[21] While some of the shorter telomere length we see in disadvantaged children might have existed from birth, in this case the newborn monkeys were randomized at birth, so these differences were purely stemming from their early experiences. Fortunately, corrective experiences later in life, like being cared for by a grandparent, can reverse some of the problems of parentless monkeys.

NURTURING CHILDREN FOR HEALTHIER TELOMERES AND BETTER EMOTION REGULATION

It is depressing to read about the maltreatment of the rat pups, or motherless monkeys. But there is a bright side to the story: The rats who were raised by nurturing mothers had healthier telomeres. Same with the monkeys. Of course, nurturing parenting is essential for human babies and children, too. Nurturing parenting can help children develop good emotion regulation, meaning that they can experience negative feelings without getting overpowered by them.[22] Think for a second, and you'll surely have no problem producing examples of adults you know who struggle to regulate their emotions. These are the people who detonate at the slightest provocation. Road rage, anyone?

Maybe you know folks at the other extreme, who find their emotions so frightening that they'd rather end a friendship than work their way through a messy disagreement. They withdraw from anything that may stir up difficult feelings—careers, friendships, even the world outside their homes. Most of us hope that our children will learn more effective means of coping.

We can teach them. From early in life, children learn to regulate their emotions through nurturing care from their parents or caregivers. The baby cries; by showing concern, the parent acts as a kind of emotional copilot, guiding the child toward an understanding of his or her emotions. By soothing the baby and tending to its needs, the parent teaches the child that it's possible to take care of feelings and to trust others. The child learns that distressing situations will eventually pass.

Fortunately for all of us who sometimes get angry in traffic or jump under the bedcovers when emotions run high, parents don't have to have perfect emotion regulation to help their children. In the reassuring words of the great English pediatrician and researcher D. W. Winnicott, they just need to be "good enough." They need to be caring, loving, and stable, with good psychological health, but they definitely don't need to be perfect. Children raised in group homes and orphanages, however, don't get anything close to good-enough parenting; they do not get the attention they need to develop normal emotional expression and regulation. They tend to have blunted emotional expression, an effect that can last throughout their lives.

The delicious act of snuggling with a baby, offering warmth, comfort, and care, has wondrous physiological effects on the child. Scientists believe that well-nurtured children learn to use their prefrontal cortex—the brain's seat of judgment—as a brake on the amygdala and its fear response. Their cortisol levels are better regulated. Put these children on a blinking, whirling kiddie ride at the state fair, or tell them that they need to take an important test, and they'll feel a healthy amount of excitement or worry. That's what stress hormones are there for—to pump us up. When the ride comes to a stop, or when they put down their pencils, the cortisol begins its retreat. They're not constantly swimming around in a flood of stress hormones.

Nurtured children also experience the delights of oxytocin, the hormone that's released when you feel close to someone. Oxytocin is a stress-busting hormone; it reduces our blood pressure and imbues us with a glowing sense of wellbeing.[23] (Women who breast-feed their children can experience the rush of oxytocin in an intense, palpable way.) Alas, the stress-buffering effect of having one's parents nearby seems to wane once children reach adolescence.[24]

A LITTLE ADVERSITY CAN BE PROTECTIVE

There is usually no upside to serious childhood adversity, just suffering and greater risk for depression and anxiety later in life. Shorter telomeres, too. Moderate adversity in childhood, however, can be healthy. Adults who report having a few—but *just* a few—adverse experiences in their youth have healthy cardiovascular responses to stress. Their hearts pumped more blood and got them ready to face the situation; in other words, they experienced a vigorous challenge response. They felt excited, invigorated—so perhaps their early experiences had given them confidence in their ability to overcome obstacles. People with *no* adverse events actually did worse. They felt more threatened, with more vasoconstriction in their peripheral arteries. (Meanwhile, those who had experienced the most severe adversity had excessive threat reactivity.)[25] We are not prescribing a dose of adversity for any child, just pointing out that it is common. If it happens *in a moderate amount, and if the child has enough support to cope with it*, there may be a benefit. Teaching children how to cope well with stress (versus protecting them from all disappointments) is the key. As Helen Keller said, "Character cannot be developed in ease and quiet. Only through experiences of trial and suffering can the soul be strengthened, vision cleared, ambition inspired and success achieved."

THE ABCS OF PARENTING VULNERABLE CHILDREN

In children who have begun their lives under traumatic circumstances, enhanced parenting techniques may help heal some of the telomere damage from early mistreatment. Mary Dozier, of the University of Delaware, has studied children who were exposed to adversity. Some lived in inadequate housing; some were neglected or witnessed or experienced domestic violence; some had parents who abused substances or who hurt each other. Dozier and her colleagues found that these children had shorter telomeres—except when their parents interacted with them in a very sensitive, responsive way.[26] To give you a sense of what this kind of parenting looks like, here's a very short assessment:

1. Your toddler bumps his head hard on a coffee table and looks at you as if ready to cry. What do you say?

- "Oh, honey, are you okay? Do you need a hug?"
- "You're okay. Hop up."
- "You shouldn't be that close to the table. Move away from there."
- You say nothing, hoping that he'll move on to something else.

2. Your child comes home from school and says that her best friend doesn't want to be friends anymore. You say:

- "I'm so sorry, honey. Do you want to talk about it?"
- "You'll have plenty of friends over time. Don't worry."
- "What did you do that made her not want to be your friend?"
- "Why don't you get on your bike and go for a ride?"

All of these answers can sound reasonable, and under individual circumstances, any of them might be. But there is only one correct response for a child who has been through trauma, and in both

306

cases, that response is the first one. Under normal conditions, it may sometimes be appropriate to help a child learn to brush off a minor bump or scrape, for example. But children who've suffered from adversity are different. They may have a harder time regulating their emotions. They still need parents to be the emotional copilot—to reassure them that the parent has noticed their troubles and can be relied upon to help soothe them. They may need this reassurance over and over and over again. It takes time, but eventually children will learn how to respond to problems in a more adaptive way. And when they are older, they will be more likely to go to their parents with issues that they are worried about.

Dozier has developed a program known as Attachment and Biobehavioral Catch-Up, or ABC, to teach this kind of exquisite responsiveness to parents of at-risk children. One group included American parents who were adopting international children. These weren't people who lacked parenting skills. They were caring and committed. But the children they were adopting were statistically much more likely to have lived in group homes, to suffer from poor emotion regulation, to have telomere damage—the whole bushel of problems that come with childhood adversity. During this program, parents are coached to *follow their child's lead.* For example, when a child starts to play a game by banging a spoon, a parent might be tempted to say, "Spoons are for stirring pudding" or "Let's count the number of times you tap the bowl." But these responses reflect the parent's agenda, not the child's. In Dozier's program, the parent would be encouraged to join in the game, or to comment on what the child is doing: "You're making a sound with your spoon and bowl!" These smooth interactions with the parent help at-risk children learn to regulate their emotions.

It's a simple intervention, but the results are dramatic. Dozier also taught ABC to a group of parents who had been reported to Child Protective Services for allegedly neglecting their children. Before the course, the children's cortisol levels had that blunted, broken response that characterizes burnout from overuse. After the parents had taken this short course, the children had a much more normal cortisol response. Their cortisol rose in the morning (a good, healthy sign that they were ready to take on the day), and declined through-out the day. This effect wasn't just temporary. It lasted for *years*.[27]

TELOMERES AND STRESS-SENSITIVE CHILDREN

Was Rose a difficult baby? Her parents smile at the question. "Rose had colic for *three years*," they say, laughing at their exaggeration as well as the kernel of truth that is behind it. Colic, in which babies cry incessantly for more than three hours a day, three days a week, gen-erally begins at around two weeks of age and usually peaks at about six weeks. Rose was colicky, all right. As a newborn, she would nurse, nap briefly, have about five minutes of peaceful time…and then begin wailing again. Despite her name, Rose was no demure flower. Her parents, desperate to calm their crying baby, would take her for walks and strolls through the neighborhood—only to have older ladies rush up to them, exclaiming, "Something must be wrong with your child! Healthy babies do not cry this way!"

Nothing was wrong. Rose was clean, fed, warm, and cared for. She was just very, very sensitive. She was quick to cry and slow to settle down for sleep or quiet—thus her parents' joke about her colic lasting for years. Small noises, like the running of the refrigerator motor, bothered her. When strangers held her, Rose would scream and try to wriggle out of their arms. As Rose got older, she wouldn't wear clothes with tags; they felt too itchy. When the family signed up for a professional photography session, Rose hid her eyes from the bright lights. And any change in her daily routine was upsetting.

Was Rose sensitive because of the way her parents raised her? Were they too indulgent of her demands? Should they have taught her a lesson by insisting, say, that Rose wear whatever clothes they picked out for her, itchy or not? We can begin to answer these questions by talking about temperament. Temperament, the set of personality traits we're born with, is like the deep cement foundation of a building. It can provide a stable undergirding, or it can make us tilt and shake in certain ways, especially during an "earthquake." We can recognize our temperament and learn to deal with it, but we can't really change our foundation. Temperament is biologically determined.

One aspect of temperament is stress sensitivity. Stress–sensitive children are more "permeable," which means that for good or for ill, their environment doesn't just bounce off them. It penetrates. These kids have bigger stress reactions to light, noise, and physical irritations. They are jolted by transitions, like going to back to school after the weekend (the "Monday effect"), or new situations, like staying at a grandparent's house overnight. They have a stronger, magnified response to shifts in their environment, even small shifts that other children might not notice. Some of these children may react by acting angry or aggressive; others may internalize their feelings, coming across as quiet or sullen. Telomeres tend to be shorter in children who internalize their emotions.[28] But when children have severe externalizing or acting-out disorders, such as attention deficit disorder with hyperactivity, and oppositional defiant disorder, their telomeres are shorter, too.[29]

Developmental pediatrician Tom Boyce has followed a group of kindergartners as they transition into their first year of school—a time that can be tough for stress-sensitive children. He and his colleagues hooked them up to sensors and then measured their physiological reactions to harmless but modestly stressful situations, like watching a scary video, having a few drops of lemon juice squirted onto their tongues, and (of course) performing one of those memory tasks. Most kids showed some signs of stress. But in a few kids, the stress responses were cranked up to their full force, both the hormonal responses and the autonomic nervous system. It was as if their bodies and brains thought the room was on fire. The bigger the stress responses, the shorter their telomeres tended to be.[30]

IS YOUR CHILD AN ORCHID?

It can all sound quite tragic. It may seem that people who are born with high-stress sensitivity have drawn the unlucky short straw—or, in this case, the short telomere. Actually, Boyce and others have found that certain environments allow stress-sensitive people to thrive, sometimes even more than their less sensitive peers.

In many studies Boyce has found that children who are especially stress sensitive do poorly when they are in large, crowded, chaotic classrooms or harsh family environments, but when they are in classrooms or families with warm, nurturing adults, they actually do better than the average child. They are less sick with colds and flu; they show fewer symptoms of depression or anxiety; they are even injured less often than other children.[31]

Boyce calls these stress-sensitive children "orchids." Without exquisite care and attention, an orchid won't bloom. Put it in the optimal conditions of a greenhouse, though, and it produces flowers of surpassing beauty. Around 20 percent of children have an orchid-like temperament. Again, it's not something that parents create. Those orchid seeds are planted long before birth.

A way to understand these "seeds" is to analyze the genetic signatures of orchid children. Children (and adults) with more variations in the genes for neurotransmitters that regulate mood, like dopamine and serotonin, tend to be more sensitive to stress. They're orchids. Those most stress sensitive, based on genetics, tend to benefit more from supportive interventions and will thrive.[32] To test whether this genetic signature affects how children's telomeres respond to adversity, a small and preliminary study looked at forty boys. Half were from stable homes; the other half were from harsh social environments characterized by poverty, unresponsive parenting, and family structures that kept changing. The boys exposed to harsh environments had shorter telomeres—but especially if they had the more stress-sensitive genes. That's the obvious disadvantage of being permeable to the environment—a rough situation is going to do deep damage. Then the boys revealed the flip side, the beauty of permeability: When they lived in stable environments, their telomeres weren't just okay. They were longer, healthier, than the telomeres of the boys without the genetic variations. This early study suggests that being sensitive and permeable may be a benefit when in a supportive environment.[33]

This is a fascinating story in personality research, and one of the hottest topics in the stress field. Sensitivity is neither a good nor a bad trait. It's just one of the cards we're dealt. It's best if we can clearly identify the card so that we can know how to play our hand. Orchid children benefit from warmth, gentle correction, and a consistent routine. They need assistance and patience as they make transitions to a new situation. As high-stress reactors, orchid children can benefit from learning the challenge response—and you can also teach them techniques like thought awareness and mindful breathing, which help them put some calming distance between themselves (their thoughts) and their active stress responses.

PARENTING TEENS FOR TELOMERE HEALTH

> **Parent:** Look at what I found underneath that mess on your desk today. Am I correct in thinking that this is an assignment for a history paper?
>
> **Teen:** I don't know.
>
> **Parent:** It's due *tomorrow*. Have you even started it yet?
>
> **Teen:** I don't know.

Parent: Answer me respectfully! Let's try again: Is this or is this not an assignment for a history paper that is due tomorrow?

Teen: I don't have to listen to this! You're just jealous because you never had fun when you were my age. You didn't know how!

Parent: You just bought yourself a grounding. You'll be staying home this Friday night.

Teen [shouting]: Go to hell!

Parent [also shouting]: AND all day Saturday!

So far we've talked about children, mostly younger ones. But what about teenagers? Parent-teen conflicts like the one above, in which an issue (like homework) is raised, fought over, but left unresolved, are common. These open-ended conflicts leave the teen with a lot of anger—and psychologists know what anger does to that cauldron of physiological responses known as stress soup. Anger heats that soup up to a rolling boil. And anger can have telomere-shortening effects, but fortunately this can be turned around through a shift in parenting style.

Gene Brody, a researcher of family studies at the University of Georgia, gives us insight into the role of parental support during the teen years, and how to bolster it. Brody tracked a group of African American teens in the impoverished rural south of the United States. It's an area where young adults leave high school only to find that there are few jobs of any kind, let alone satisfying jobs, and few resources to help them make the transition to adult life. Alcohol use in particular is high. Brody recruited a group of these teens for his Adults in the Making program, in which teens are given emotional support and job advice. The instructors also provide strategies for handling racism. The teens' parents are included in the program, too—they're taught to tell their child in clear, vigorous terms to stay away from drugs and alcohol, for example. They have six classes where parents and teens learn skills in separate groups and then practice them together at the end. Half the teens did not get the classes. Five years later, Brody measured their telomeres. First of all, having unsupportive parenting—lots of arguments and little emotional support—was associated with shorter telomere length and more substance use five years later. However, among this vulnerable group, the teens that had received the supportive intervention had longer telomeres compared to teens who hadn't. This effect is partly explained by the teens feeling less angry.[34]

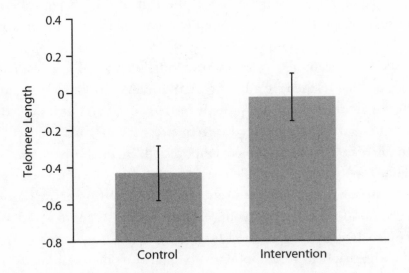

Figure 28: Family Resilience Classes and Telomeres. Among the teenagers whose parents showed very unsupportive parenting, those who were in the supportive intervention group had significantly longer telomeres five years later. (This is after adjusting for factors such as social status, stressful events, smoking, alcohol use, and body mass index.)[35]

Brody's study looked at teens in a very particular setting and at a certain income level. But his findings provide food for thought for all of us. No matter where they live, and no matter how rich or poor, *all* children's brains and bodies are undergoing tremendous changes during adolescence. It's common for teens to follow a jagged path for a while, especially because the teen brain experiences risk differently. They tend to react to threat as a thrill; when they take risks, they feel good.[36] The same behaviors are, naturally, terrifying to the more seasoned adults in their lives. Cue the parental worries, dead-of-night ruminations, and fears that explode into fights between parent and teen. A few conflicts are probably unavoidable. But when conflicts are constant, or when the tension becomes so toxic that it pollutes the air of the household, teens can become angry and rebellious. Or depressed and anxious, if they're the type to drive their feelings underground. The Renewal Lab at the end of this chapter offers a few suggestions for staying attuned to teens when they are in a difficult, hyperreactive mode.

We have been talking about how to help children heal the telomere damage caused by adversity. Early intervention, support, and emotional attunement can provide buffers for at-risk children. But you may have had prolonged, severe stress in early life yourself. If you grew up in a dangerous neighborhood, in an abusive home, or if your family had to struggle just to get food and shelter, your telomeres may have experienced some damage. Use this knowledge as motivation to take care of your telomeres now. Recognize old patterns, such as turning to food for comfort. You have more control over what happens to you now that you are an adult. And now you know how to protect the base pairs of telomeres you have left. You may especially want to take advantage of techniques that help soothe the stress response. By becoming less stress reactive, you will protect your telomeres—and there is a bonus. You will also be calmer and stronger for the children (and other loved ones) who are in your life.

TELOMERE TIPS

- Severe childhood trauma is linked to shorter telomeres. Trauma can also reverberate into adulthood in the form of poor health behaviors and relationship difficulties, which may continue to shorten telomeres. If you suffered severe childhood adversity, you can take steps now to buffer its effects on your wellbeing and telomeres.

■ Although severe childhood adversity can be damaging, moderate childhood stress may actually be healthy, provided that the child has enough support during the stressful time.

■ Parents can support their young children's telomeres by practicing warm, nurturing attunement. This responsiveness is especially important for children who have already experienced trauma or who are born with the sensitive "orchid" temperament.

RENEWAL LAB

WEAPONS OF MASS DISTRACTION

The ABC program teaches parents to avoid unresponsive behaviors, including something that almost all of us are guilty of: distraction. No matter what a child's situation or temperament, being connected to a screen means we're not connected to the child. And it's easier to be distracted than you might think. When a cell phone is present on a nearby table, people engage in conversation that is more shallow, and their attention is more divided.[37] Digital conversations limit the opportunity for full empathy and connection. No wonder the writer Pico Iyer refers to smartphones as "weapons of mass distraction."

This Renewal Lab invites you to engage with the children in your life *without* the interference of screens. *See if you can spend twenty minutes talking to a child, playing a game, or just enjoying his or her presence, without a phone or tablet computer nearby. Limit your children's screen time as well. Make it intentional—sometimes naming something gives it a lot more power and makes it more effective. Though your child may resist this screen-free drive or mealtime, he or she may also welcome it, albeit secretly. Decide on a few critical screen-free times such as meals, the car ride to and from school, and the first half hour after coming into the house after being away (when attention should be focused on reconnecting with the family). If the screen-free time is a clear rule, you won't need to get into complex negotiations every day. For tips on how to "outsmart*

smart screens" and limit your child's use, Harvard's Prevention Research Center has a free guide for parents: http://www.hsph .harvard.edu/prc/2015/01/07/outsmarting-the-smart-screens/. Your family can also participate in Screen-Free Week, a campaign hosted by the Campaign for a Commercial Free Childhood each spring (http://www.screenfree.org/).

TUNING IN TO YOUR CHILD

Vulnerable children need tremendous sensitivity and parental attunement. You can soothe some of their frustration by tuning in to their feelings. Homework, for example, is a common stressor. Kids get upset about the homework itself, and they can also get irritated with their parents when they try to help. Daniel Siegel, author of *The Whole-Brain Child* and coauthor of *Brainstorm*, offers ways to attune especially when riding waves of high emotions. He explains that parents can't help their child manage homework (or any other stressful activity) until they acknowledge and empathize with the child's feelings.

So the next time your child is under stress, try saying something that acknowledges his or her feelings, such as "You look frustrated." You can also help your child identify his or her feelings, since labeling feelings and putting together a story of what happened turns down the volume on emotions. Siegel calls this strategy "name it to tame it." You can say things like, "Wow, that seemed like a tough situation. What was that like for you? How are you feeling?" If you want to reach the child's rational thinking, you have to meet the child emotionally first, with empathy.[38] Siegel calls this "connect and redirect."

DON'T OVERREACT TO YOUR REACTIVE TEEN

Don't let your teen's emotional, thrill-seeking brain pull you into an escalating conflict. If your teen is ranting at you, you have options other than your automatic one...reacting. An argument cannot escalate if you are not part of it. Sometimes it helps to say you need a time-out for yourself, a small dose of time and space in a different place. Given the quick half-life of emotions, your child's emotions (and yours) will most likely subside, and a conversation can resume with both sides of the brain working.

In the heat of the moment, you may remind yourself that although adolescents may look like grown-ups on the outside, they are still children on the inside. They need you to be clear and steady, not to get tangled up in their drama. Remind yourself *you're* the one in the room with the adult brain—and that you have the power to remain calm and to avoid escalating the argument. Also, in calm moments, be curious. Rather than tell your teen what to do, ask him or her questions.

BE A MODEL OF LOVING ATTACHMENT

A loving relationship with your partner is not only precious but a tool for better parenting. One study tracked children's reactions to their parents' daily interactions over three months. The study examined how much emotional resonance or mirroring children had to their parents' interactions. When the parents showed affection to each other and the children felt more positive affect, the children tended to have longer telomeres. Conversely, when the parents had conflict, and the child responded with negative emotions, the child tended to have shorter telomeres.[39] So remember that emotions are permeable, and especially for sensitive children. Consider warming up your family environment and showing your affection. This is hard to do if angry emotions are running high! But by showing love for your partner, you may also be promoting wellbeing in your child (and maybe his or her telomeres, too).

Conclusion
Entwined: Our Cellular Legacy

A human being is part of the whole, called by us "Universe,"
a part limited in time and space. He experiences himself, his
thoughts and feelings, as something separated from the rest—
a kind of optical delusion of his consciousness. This delusion
is a kind of prison for us, restricting us to our personal desires
and to affection for a few persons nearest to us. Our task
must be to free ourselves from this prison by widening our
circle of compassion to embrace all living creatures and the
whole of nature in its beauty. Nobody is able to achieve this
completely, but the striving for such an achievement is in itself
a part of the liberation and a foundation for inner security.
　　　　—Albert Einstein, as quoted in the *New York Times*,
　　　　　　　　　　　　　　　　　　　　　March 29, 1972

A long life of good health and wellbeing is our hope for you. Lifestyle, mental health, and environment all contribute significantly to physical health— that is not new. What is new here is that telomeres are impacted by these factors, and thus quantify their contribution in a clear and powerful way. The fact that we see the transgenerational impacts of these influences makes the message of telomeres all the more urgent. Our genes are like computer hardware; we cannot change them. Our epigenome, of which telomeres are a part, is like software, which requires programming. We are the programmers of the epigenome. To some extent, we control the chemical signals that orchestrate the changes. Our telomeres are responsive, listening, calibrating to the current circumstances in the world. Together we can improve the programming code.

The preceding pages have been full of our best suggestions, gleaned from hundreds of studies, about how to protect your precious telomeres. You've seen how telomeres are affected by your mind. You've seen how they're shaped by your habits of movement, by the quality and length of your sleep, and by the foods you eat. Telomeres are also affected by the world beyond the mind and body—because your neighborhood and relationships foster a sense of safety that can shape your telomere health.

Unlike humans, telomeres do not make judgments. They are objective and unbiased. Their reaction to their environment is quantifiable, down to their base pairs. This makes them an ideal index for measuring the effects of our internal and external environment on our health. If we listen to what they have to tell us, telomeres provide insight into how we can prevent premature cellular aging and promote our healthspan. But as it turns out, the story of healthspan is also the story of what a beautiful life and world can look like. What is good for our own telomeres is also good for our children, our community, and people around the globe.

TELOMERES SOUND THE ALARM

Telomeres teach us that from the earliest days of life, severe stress and adversity reverberate all the way into adulthood, setting up our youngest generation for a life marred by higher likelihood of early chronic disease. In particular, we've learned that childhood exposure to stressors like violence, trauma, abuse, and socioeconomic hardship have all been linked to shorter telomeres in adulthood. The damage may begin even before the child is born; high maternal stress may be transmitted to the developing fetus in the form of shorter telomeres.

This early imprint of stress on telomeres is an alarm bell. We call for policymakers to add a new phrase, **societal stress reduction**, to the vocabulary of public health. We're not speaking here of exercise or yoga classes, though these are helpful to many people. We're talking about broad social policies that have the goal of buffering the ubiquitous socioenvironmental and economic chronic stressors faced by so many.

The worst stressors—exposure to violence, trauma, abuse, and mental illness—are shaped by a surprising factor: the level of income inequality in a region. For example, countries with the biggest gap between their richest citizens and their poorest have the worst health and the most violence. As you can see from figure 29, these countries also have the highest rates of depression, anxiety, and schizophrenia.[1]

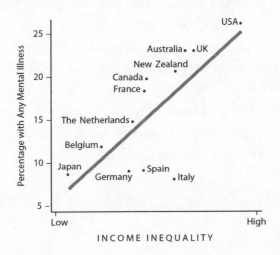

Figure 29: Income Inequality and Mental Health. A large body of research has shown that income inequality in regions and countries is associated with worse behavior (less trust, more violence, drug abuse) and worse health for all, whether it's physical or mental health. Kate Pickett and Richard Wilkinson have summarized this massive set of research findings,[2] and here they show relations with mental health. In this data set, Japan has the lowest inequality and the lowest rate of mental illness, whereas the United States has the highest of each.

A substantial number of studies have demonstrated this relationship. And it's not just the poor who suffer from the gap. Everyone in these stratified societies is at a higher risk for impaired mental and physical health—and the more unequal the society, the lower the child wellbeing. You see this effect among rich and poor states in the United States. The inequality gap has been widening such that, in the United States, the top 3 percent in the distribution owns 50 percent of the wealth[3] (no wonder the United States has the biggest gap among the rich countries). Tellingly, Sweden, which has the lowest income gap of all countries, also has the highest wellbeing, including the wellbeing of children. But it's also one of the countries with the fastest growing inequality and decreasing child wellbeing (due to a reduction of the redistributive effect of Sweden's tax and benefit system).[4]

We believe that the income gap drives the difference between the likelihood of healthy, long, stable telomeres in old age and mere stumps of short ones from aged, senescent cells. This gap represents excessive social stress, competitive stress, and the sickness in societies that leads to an early and prolonged diseasespan for both rich and poor. An essential element of societal stress reduction is the narrowing of this enormous gap. Understanding how we are interconnected is the fuel that will drive this work.

INTERCONNECTEDNESS AT ALL LEVELS

We are connected to one another and to all living things at all levels from the macro to the micro, from the societal to the cellular. The separation we all feel, as if we are each on a path alone, is an illusion. The reality is that we all share much more than we can ever comprehend, both in mind and body. We are deeply interconnected with each other and nature in phenomenal ways.

Within our body and cells, we're connected to other living organisms. Our bodies are made up of eukaryotic cells. It's thought that about a billion and a half years ago, long before humans evolved, the single eukaryotic cell swallowed up bacterial organisms that lived together interdependently as one cell. The mitochondria that live in our cells today are the legacy of these bacteria and this interdependence. We're symbiotic creatures.

Inside our bodies, we carry around a shared part of the outside world. Roughly two to three pounds of human weight is made up of other beings: microbes. Microbes live as complex communities within our gut and on our skin. Far from being our sworn enemies, they keep us balanced. Without these colonies of microbes, our immune system would be weak and underdeveloped; they send signals to our brain and can make us depressed when they are out of balance. And it works the other way, too—when we are feeling depressed or stressed, we are affecting our microbiota, impairing their balanced state, and impairing our mitochondria.[5]

Humans are increasingly interconnected to one another—from technology to financial markets, to media and social network groups. We are always embedded in a social culture, and our thoughts and feelings are shaped by our immediate social and physical surroundings.[6] Our perceptions of how supported and connected we are matter for our health. This has always been true, but now those connections are becoming both broader and tighter. A global broadband will soon encircle the world, allowing everyone on the planet to be connected via the Internet in a highly affordable way. On any random day last year, one out of seven people worldwide had logged on to Facebook.[7] This growing interconnectedness opens up further opportunities to unite around the issues that are most significant to us.

We're also sharing the same physical environment. Pollution on one side of the world can travel to the other, blowing in the wind or floating in the water. Together, we are heating up our globe, and we are all affected by it. It's another sign of how we are connected, an urgent reminder that our daily behavior matters.

Finally, we're connected from generation to generation. We now know that telomeres are transmitted through the generations. The disadvantaged unknowingly pass on that disadvantage—through economic and social problems, but also likely through shorter telomeres and other epigenetic paths. In this way, telomeres are our message to our future society. Worse, children are being exposed to toxic stress at epidemic levels, leaving them with shorter telomeres and premature cellular aging. As John F. Kennedy reminded us, "Children are the living messages we send to a time we will not see." We do not want that message to include early chronic disease. This is why it's so important to cultivate our inborn sense of compassion. We must rewrite that message.

THE LIVING MESSAGES

Telomere science has grown into a clarion call. It tells us that social stressors, especially as they affect children, will result in exponentially higher costs down the line—costs that are personal, physical, social, and economic. You can respond to that call by, first, taking good care of yourself.

The call doesn't end there. Now that you know how to protect your telomeres, we want to issue you a friendly challenge. What will you do with your many decades of brimming good health? A long healthspan makes a vital, energetic life more possible, and that vitality can ripple outward, allowing us to spend some of our time creating conditions for better health and wellbeing in other people.

We can't eliminate stress and adversity, of course, but there are ways to relieve some of the extreme pressure on the most vulnerable populations. We've told you about painful aspects of some people's lives, but that's just one aspect of their lives. Robin Huiras, the woman with an inherited telomere disorder, who has helped recruit some of the best minds in telomere science to write the first clinical handbook for treatment of telomere disorders, is helping to alleviate suffering. Peter, the medical researcher who struggles with a brain bent on overeating, travels around the world on medical missions to underserved people and has filled his life with purpose and contribution. Tim Parrish, the man who grew up in a racist community in Louisiana, writes and speaks about this painful subject, risking his own comfort to help us face our prejudices more effectively.

What is your cellular legacy? Each of us has a time-limited opportunity to leave a legacy. Just as your body is a community of individual but mutually dependent cells, we are a world of interdependent people. We all have an impact on the world, whether we realize it or not. Large changes, such as implementing policies for societal stress reduction, are vital. Small changes are important, too. How we interact with other people shapes their feelings and sense of trust. *Every day, each of us has the chance to positively influence the life of another person.*

The story of telomeres can inspire our determination to elevate our collective health. Helping to change our communities and shared environment gives us that vital sense of mission and purpose, which itself may improve our telomere maintenance.

The foundation for a new understanding of health in our society is not about "me" but "we." Redefining healthy aging is not just about accepting gray hair and focusing on inner health; it's also about our connections with others and building safe, trusting communities. Telomere science offers molecular proof of the importance of societal health to our individual wellbeing. We now have a way to index and measure the interventions we create to improve that health. Let's get started.

THE TELOMERE MANIFESTO

Your cellular health is reflected in the wellbeing of your mind, body, and community. Here are the elements of telomere maintenance that we believe to be the most crucial for a healthier world:

Mind Your Telomeres

- Evaluate sources of persistent, intense stress. What can you change?
- Transform a threat to a challenge appraisal.
- Become more self-compassionate and compassionate to others.
- Take up a restorative activity.
- *Practice thought awareness and mindful attention. Awareness opens doors to wellbeing.*

Maintain Your Telomeres

- Be active.
- Develop a sleep ritual for more restorative and longer sleep.
- Eat mindfully to reduce overeating and ride out cravings.
- Choose telomere-healthy foods—whole foods, omega-3s, skip the bacon.

Connect Your Telomeres

- Make room for connection: Disconnect from screens for part of the day.
- Cultivate a few good, close relationships.
- Provide children quality attention and the right amount of "good stress."
- Cultivate your neighborhood social capital. Help strangers.
- Seek green. Spend time in nature.
- *Mindful attention to other people allows connections to bloom. Attention is your gift to give.*

Create Telomere Health in Your Community and the World

- Improve prenatal care.
- Protect children from violence and other traumas that damage telomeres.
- Reduce inequality.
- Clean up local and global toxins.
- Improve food policies so that everyone has access to fresh, healthy, affordable food.

The future health of our society is being shaped right now, and we can measure part of that future in telomere base pairs.

Acknowledgments

We could not have written this book without drawing upon the decades of hard work of many scientists, and we thank them all for their contributions to our understanding of telomeres, human aging, and behavior, even though we couldn't reference each important contribution of our colleagues. We thank the innumerable scientific collaborators and students with whom we have worked during the past few decades; our gratitude to each of you is bottomless. Our research could not have transpired without you. We are both especially indebted to Dr. Jue Lin, PhD, who has worked tirelessly and with great talent for over ten years on all of our human telomere studies. Jue has performed tens of thousands of meticulous telomere-length and telomerase measurements for these studies, and has served as an exemplar of a translational researcher, working at all levels, from lab bench to community.

We would like to acknowledge the following people who have contributed to this book in various important ways, through enlightening discussions, providing perspectives on the book, or serving as an inspiration or support to our work. Any mistakes in the content, however, are entirely our own. We extend our deepest gratitude to: Nancy Adler, Mary Armanios, Ozlum Ayduk, Albert Bandura, James Baraz, Roger Barnett, Susan Bauer-Wu, Peter and Allison Baumann, Petra Boukamp, Gene Brody, Kelly Brownell, Judy Campisi,

Laura Carstensen, Steve Cole, Mark Coleman, David Creswell, Alexandra Croswell, Susan Czaikowski, James Doty, Mary Dozier, Rita Effros, Sharon Epel, Michael Fenech, Howard Friedman, Susan Folkman, Julia Getzelman, Roshi Joan Halifax, Rick Hecht, Jeannette Ickovics, Michael Irwin, Roger Janke, Oliver John, Jon Kabat-Zinn, Will and Teresa Kabat-Zinn, Noa Kageyama, Erik Kahn, Alan Kazdin, Lynn Kutler, Barbara Laraia, Cindy Leung, Becca Levy, Andrea Lieberstein, Robert Lustig, Frank Mars, Pamela Mars, Ashley Mason, Thea Mauro, Wendy Mendes, Bruce McEwen, Synthia Mellon, Rachel Morello-Frosch, Judy Moskowitz, Belinda Needham, Kristen Neff, Charles Nelson, Lisbeth Nielsen, Jason Ong, Dean Ornish, Bernard and Barbro Osher, Alexsis de Raadt St. James, Judith Rodin, Brenda Penninx, Ruben Perczek, Kate Pickett, Stephen Porges, Aric Prather, Eli Puterman, Robert Sapolsky, Cliff Saron, Michael Scheier, Zindel Segal, Daichi Shimbo, Dan Siegel, Felipe Sierra, the late Richard Suzman, Shanon Squires, Matthew State, Janet Tomiyama, Bert Uchino, Pathik Wadhwa, Mike Weiner, Christian Werner, Darrah Westrup, Mary Whooley, Jay Williams, Redford Williams, Janet Wojcicki, Owen Wolkowitz, Phil Zimbardo, and Ami Zota. A big thanks to Aging, Metabolism, and Emotions (AME) lab members, and especially to Alison Hartman, Amanda Gilbert, and Michael Coccia, for support on various aspects of the book. We thank Coleen Patterson of Coleen Patterson Design for her inspired illustrations, and amazing transference of images from our heads to this book.

We thank Thea Singer for covering the telomere-stress connection so beautifully in her book *Stress Less* (Hudson Street Press, 2010). We also thank the dedicated readers of our book focus group who gave us their Sunday afternoons and invaluable input: Michael Acree, Diane Ashcroft, Elizabeth Brancato, Miles Braun, Amanda Burrowes, Cheryl Church, Larry Cowan, Joanne Delmonico, Tru Dunham, Ndifreke Ekaette, Emele Faifua, Jeff Fellows, Ann Harvie, Kim Jackson, Kristina Jones, Carole Katz, Jacob Kuyser, Visa Lakshi, Larissa Lodzinski, Alisa Mallari, Chloe Martin, Heather McCausland, Marla Morgan,

Debbie Mueller, Michelle Nanton, Erica "Blissa" Nizzoli, Sharon Nolan, Lance Odland, Beth Peterson, Pamela Porter, Fernanda Raiti, Karin Sharma, Cori Smithen, Sister Rosemarie Stevens, Jennifer Taggart, Roslyn Thomas, Julie Uhernik, and Michael Worden. Thanks to Andrew Mumm of Idea Architects for his wizardry and patience connecting us across geographic and technical challenges.

We'd also like to thank the people who generously talked with us about their personal experiences, some anonymously, some named below. We weren't able to incorporate every single one of the wonderful stories we heard, but throughout the writing process the spirit of all those stories has informed and profoundly moved us. We are indebted to Cory Brundage, Robin Huiras, Sean Johnston, Lisa Louis, Siobhan Mark, Leigh Anne Naas, Chris Nagel, Siobhan O'Brien, Tim Parrish, Abby McQueeney Penamonte, Rene Hicks Schleicher, Maria Lang Slocum, Rod E. Smith, and Thulani Smith.

We extend *tremendous* thanks to Leigh Ann Hirschman, of Hirschman Literary Services, our collaborative writer. Her writing and depth of editorial experience helped make this book as readable as it is. She was a pleasure to work with: joining our immersion in the world of telomere science, ever patient with our bringing in the constant flow of new studies that entered the scientific literature as we wrote, and a balanced and guiding voice when at times we thought we would never emerge from the thickets of research.

We are also very grateful to our editor, Karen Murgolo of Grand Central Publishing, for her faith in this book and her expertise, time, and care in every decision needed throughout this process. We felt so fortunate to have benefitted from her wisdom and patience.

We have deep gratitude to Doug Abrams of Idea Architects. It was Doug who first saw the need for a book that we could not yet see. We thank him for his dedication and for his wonderful and wise curation as a developmental editor. And for making what could have been taxing to our telomeric base pairs both a delightful process and the grounding of enduring friendship.

Lastly, we are so grateful to our families (nuclear and extended) for their loving support and enthusiasm during the many seasons of the writing process, and the many more seasons that laid the scientific foundation for it.

We are also grateful for the opportunity to share this work with you, the readers, and sincerely hope that this work promotes your wellbeing and healthspan.

Information about Commercial Telomere Tests

If you'd like to estimate your telomere health, you can take the self-test on page 161. You can also take a commercial company test to determine telomere length. But should you? You don't need to have your own lungs biopsied in order to make the wise decision to stop smoking! Many of you would probably perform the same restorative activities in life whether you have a telomere test or not.

We wondered how people would react to learning the results of telomere tests. If a person learns he or she has short telomeres, for example, would that knowledge be depressing? So we tested volunteers and told them their results. Then we followed up to ask about their reactions. Most were neutral to positive, and none was very negative. But those who were short did experience some distressing thoughts about that in the ensuing months. Telomere testing is a personal decision. Only you can decide if knowing your length will benefit you. Imagine if you learn your telomeres are short— is that more motivating to you than upsetting? Learning that your telomeres are short is like seeing the "check engine" light on a dashboard; it's usually just a sign that you need to take a closer look at your health and your habits and step up your efforts.

We're often asked if we've had our own telomeres measured:

I (Liz) have, out of curiosity. My results were reassuringly good, but I always keep in mind that telomere length is a statistical indication of health, not an absolute predictor of the future.

I (Elissa) haven't had my telomeres measured yet. I would rather not know definitively if my telomeres are short. I try to engage in the life practices good for telomeres on an AMAP (as much as possible) basis, given this busy life. Telomere length trajectories over time will be more valuable than single checks. They tell us something unique about a cell's potential to replicate that no one indicator can. However, they are just one marker. It is likely that algorithms including many biomarkers and health status variables will be more beneficial for personal use once they are better developed. When the measures have more predictive value for individuals and are easier to get repeatedly, I will be more interested in getting testing done.

As of this writing, only a few commercial companies offer telomere testing.

We do not have any knowledge about—or control over—the accuracy and reliability of telomere length measurements performed by these commercial entities. Because these companies change rapidly, we list the details on our book website. At this writing, testing costs anywhere from around $100 to $500.

A few caveats: Telomere testing is an unregulated business, so there is no government agency checking whether for-profit companies are using methods and values that are accurate, or whether what they tell you about your risks is accurate. It can be interesting to learn the results of a telomere test, but we caution everyone that telomeres do not necessarily predict the future. Again, it's like smoking. Smoking does not guarantee that you'll get a lung disease, and not smoking does not guarantee that you will stay free of disease. But the statistics on smoking are in, and the message is clear: the more you smoke, the greater your chances of getting emphysema, cancer, and other serious health problems. There are plenty of good reasons to quit—or better still, not to smoke at all. In the same way, the countless studies on the relationship between telomere length and human health and disease have given us the data we need to create guidelines for keeping your telomeres (and therefore you) healthier. You may enjoy knowing your telomere length, but you don't need that information to prevent premature cellular aging.

Notes

Authors' Note: Why We Wrote This Book

1. "Oldest Person Ever," Guinness World Records, http://www.guin
nessworldrecords.com/world-records/oldest-person, accessed March 3,
2016.
2. Whitney, C. R., "Jeanne Calment, World's Elder, Dies at 122," *New York
Times*, August 5, 1997, http://www.nytimes.com/1997/08/05/world
/jeanne-calment-world-s-elder-dies-at-122.html, accessed March 3, 2016.
3. Blackburn, E., E. Epel, and J. Lin, "Human Telomere Biology: A Contributory and Interactive Factor in Aging, Disease Risks, and Protection," *Science* 350, no. 6265 (December 4, 2015): 1193–98.

Introduction: A Tale of Two Telomeres

1. Bray, G. A. "From Farm to Fat Cell: Why Aren't We All Fat?" *Metabolism* 64, no. 3 (March 2015):349–353, doi:10.1016/j.metabol.2014.09.012,
Epub 2014 Oct 22, PMID: 25554523, p. 350.
2. Christensen, K., G. Doblhammer, R. Rau, and J. W. Vaupel, "Ageing
Populations: The Challenges Ahead," *Lancet* 374, no. 9696 (October 3,
2009): 1196–1208, doi:10.1016/S0140-6736(09)61460-4.
3. United Kingdom, Office for National Statistics, "One Third of Babies
Born in 2013 Are Expected to Live to 100," December 11, 2013, The
National Archive, http://www.ons.gov.uk/ons/rel/lifetables/historic
-and-projected-data-from-the-period-and-cohort-life-tables/2012
-based/sty-babies-living-to-100.html, accessed November 30, 2015.
4. Bateson, M., "Cumulative Stress in Research Animals: Telomere Attrition as a Biomarker in a Welfare Context?" *BioEssays* 38, no. 2 (February
2016): 201–12, doi:10.1002/bies.201500127.

5. Epel, E., E. Puterman, J. Lin, E. Blackburn, A. Lazaro, and W. Mendes, "Wandering Minds and Aging Cells," *Clinical Psychological Science* 1, no. 1 (January 2013): 75–83, doi:10.1177/2167702612460234.

6. Carlson, L. E., et al., "Mindfulness-Based Cancer Recovery and Supportive-Expressive Therapy Maintain Telomere Length Relative to Controls in Distressed Breast Cancer Survivors." *Cancer* 121, no. 3 (February 1, 2015): 476–84, doi:10.1002/cncr.29063.

Chapter One: How Prematurely Aging Cells Make You Look, Feel, and Act Old

1. Epel, E. S., and G. J. Lithgow, "Stress Biology and Aging Mechanisms: Toward Understanding the Deep Connection Between Adaptation to Stress and Longevity," *Journals of Gerontology, Series A: Biological Sciences and Medical Sciences* 69 Suppl. 1 (June 2014): S10–16, doi:10.1093/gerona/glu055.

2. Baker, D. J., et al., "Clearance of p16Ink4a-positive Senescent Cells Delays Ageing-Associated Disorders," *Nature* 479, no. 7372 (November 2, 2011): 232–36, doi:10.1038/nature10600.

3. Krunic, D., et al., "Tissue Context-Activated Telomerase in Human Epidermis Correlates with Little Age-Dependent Telomere Loss," *Biochimica et Biophysica Acta* 1792, no. 4 (April 2009): 297–308, doi:10.1016/j.bbadis.2009.02.005.

4. Rinnerthaler, M., M. K. Streubel, J. Bischof, and K. Richter, "Skin Aging, Gene Expression and Calcium," *Experimental Gerontology* 68 (August 2015): 59–65, doi:10.1016/j.exger.2014.09.015.

5. Dekker, P., et al., "Stress-Induced Responses of Human Skin Fibroblasts in Vitro Reflect Human Longevity," *Aging Cell* 8, no. 5 (September 2009): 595–603, doi:10.1111/j.1474-9726.2009.00506.x; and Dekker, P., et al., "Relation between Maximum Replicative Capacity and Oxidative Stress-Induced Responses in Human Skin Fibroblasts in Vitro," *Journals of Gerontology, Series A: Biological Sciences and Medical Sciences* 66, no. 1 (January 2011): 45–50, doi:10.1093/gerona/glq159.

6. Gilchrest, B. A., M. S. Eller, and M. Yaar, "Telomere-Mediated Effects on Melanogenesis and Skin Aging," *Journal of Investigative Dermatology Symposium Proceedings* 14, no. 1 (August 2009): 25–31, doi:10.1038/jidsymp.2009.9.

7. Kassem, M., and P. J. Marie, "Senescence-Associated Intrinsic Mechanisms of Osteoblast Dysfunctions," *Aging Cell* 10, no. 2 (April 2011): 191–97, doi:10.1111/j.1474-9726.2011.00669.x.

8. Brennan, T. A., et al., "Mouse Models of Telomere Dysfunction Phenocopy Skeletal Changes Found in Human Age-Related Osteoporosis,"

Disease Models and Mechanisms 7, no. 5 (May 2014): 583–92, doi:10.1242/dmm.014928.

9. Inomata, K., et al., "Genotoxic Stress Abrogates Renewal of Melanocyte Stem Cells by Triggering Their Differentiation," *Cell* 137, no. 6 (June 12, 2009): 1088–99, doi:10.1016/j.cell.2009.03.037.

10. Jaskelioff, M., et al., "Telomerase Reactivation Reverses Tissue Degeneration in Aged Telomerase-Deficient Mice," *Nature* 469, no. 7328 (January 6, 2011): 102–6, doi:10.1038/nature09603.

11. Panhard, S., I. Lozano, and G. Loussouam, "Greying of the Human Hair: A Worldwide Survey, Revisiting the '50' Rule of Thumb," *British Journal of Dermatology* 167, no. 4 (October 2012): 865–73, doi:10.1111/j.1365-2133.2012.11095.x.

12. Christensen, K., et al., "Perceived Age as Clinically Useful Biomarker of Ageing: Cohort Study," *BMJ* 339 (December 2009): b5262.

13. Noordam, R., et al., "Cortisol Serum Levels in Familial Longevity and Perceived Age: The Leiden Longevity Study," *Psychoneuroendocrinology* 37, no. 10 (October 2012): 1669–75; Noordam, R., et al., "High Serum Glucose Levels Are Associated with a Higher Perceived Age," *Age (Dordrecht, Netherlands)* 35, no. 1 (February 2013): 189–95, doi:10.1007/s11357-011-9339-9; and Kido, M., et al., "Perceived Age of Facial Features Is a Significant Diagnosis Criterion for Age-Related Carotid Atherosclerosis in Japanese Subjects: J-SHIPP Study," *Geriatrics and Gerontology International* 12, no. 4 (October 2012): 733-40, doi:10.1111/j.1447-0594.2011.00824.x.

14. Codd, V., et al., "Identification of Seven Loci Affecting Mean Telomere Length and Their Association with Disease," *Nature Genetics* 45, no. 4 (April 2013): 422–27, doi:10.1038/ng.2528.

15. Haycock, P. C., et al., "Leucocyte Telomere Length and Risk of Cardiovascular Disease: Systematic Review and Meta-analysis," *BMJ* 349 (July 8, 2014): g4227, doi:10.1136/bmj.g4227.

16. Yaffe, K., et al., "Telomere Length and Cognitive Function in Community-Dwelling Elders: Findings from the Health ABC Study," *Neurobiology of Aging* 32, no. 11 (November 2011): 2055–60, doi:10.1016/j.neurobiolaging.2009.12.006.

17. Cohen-Manheim, I., et al., "Increased Attrition of Leukocyte Telomere Length in Young Adults Is Associated with Poorer Cognitive Function in Midlife," *European Journal of Epidemiology* 31, no. 2 (February 2016), doi:10.1007/s10654-015-0051-4.

18. King, K. S., et al., "Effect of Leukocyte Telomere Length on Total and Regional Brain Volumes in a Large Population-Based Cohort,"

JAMA Neurology 71, no. 10 (October 2014): 1247–54, doi:10.1001/jamaneurol.2014.1926.

19. Honig, L. S., et al., "Shorter Telomeres Are Associated with Mortality in Those with APOE Epsilon4 and Dementia," *Annals of Neurology* 60, no. 2 (August 2006): 181–87, doi:10.1002/ana.20894.

20. Zhan, Y., et al., "Telomere Length Shortening and Alzheimer Disease—A Mendelian Randomization Study," *JAMA Neurology* 72, no. 10 (October 2015): 1202–03, doi:10.1001/jamaneurol.2015.1513.

21. If you would like, you can contribute to studies on brain aging and disease without having to get your brain scanned, or even show up in person. Dr. Mike Weiner, a noted researcher at UCSF who leads the largest cohort study of Alzheimer's disease worldwide, developed the online Brain Health Registry. By joining the Brain Health Registry you answer questionnaires and take online cognitive tests. We are helping him study the effects of stress on brain aging. You can find the registry at http://www.brainhealthregistry.org/

22. Ward, R. A., "How Old Am I? Perceived Age in Middle and Later Life," *International Journal of Aging and Human Development* 71, no. 3 (2010): 167–84.

23. Ibid.

24. Levy, B., "Stereotype Embodiment: A Psychosocial Approach to Aging," *Current Directions in Psychological Science* 18, vol. 6 (December 1, 2009): 332–36.

25. Levy, B. R., et al., "Association Between Positive Age Stereotypes and Recovery from Disability in Older Persons," *JAMA* 308, no. 19 (November 21, 2012): 1972–73, doi:10.1001/jama.2012.14541; Levy, B. R., A. B. Zonderman, M. D. Slade, and L. Ferrucci, "Age Stereotypes Held Earlier in Life Predict Cardiovascular Events in Later Life," *Psychological Science* 20, no. 3 (March 2009): 296–98, doi:10.1111/j.1467-9280.2009.02298.x.

26. Haslam, C., et al., "'When the Age Is In, the Wit Is Out': Age-Related Self-Categorization and Deficit Expectations Reduce Performance on Clinical Tests Used in Dementia Assessment," *Psychology and Aging* 27, no. 3 (April 2012): 778784, doi:10.1037/a0027754.

27. Levy, B. R., S. V. Kasl, and T. M. Gill, "Image of Aging Scale," *Perceptual and Motor Skills* 99, no. 1 (August 2004): 208–10.

28. Ersner-Hershfield, H., J. A. Mikels, S. J. Sullivan, and L. L. Carstensen, "Poignancy: Mixed Emotional Experience in the Face of Meaningful Endings," *Journal of Personality and Social Psychology* 94, no. 1 (January 2008): 158–67.

29. Hershfield, H. E., S. Scheibe, T. L. Sims, and L. L. Carstensen, "When Feeling Bad Can Be Good: Mixed Emotions Benefit Physical Health Across Adulthood," *Social Psychological and Personality Science* 4, no.1 (January 2013): 54–61.

30. Levy, B. R., J. M. Hausdorff, R. Hencke, and J. Y. Wei, "Reducing Cardiovascular Stress with Positive Self-Stereotypes of Aging," *Journals of Gerontology, Series B: Psychological Sciences and Social Sciences* 55, no. 4 (July 2000): P205–13.

31. Levy, B. R., M. D. Slade, S. R. Kunkel, and S. V. Kasl, "Longevity Increased by Positive Self-Perceptions of Aging," *Journal of Personal and Social Psychology* 83, no. 2 (August 2002): 261–70.

Chapter Two: The Power of Long Telomeres

1. Lapham, K. et al., "Automated Assay of Telomere Length Measurement and Informatics for 100,000 Subjects in the Genetic Epidemiology Research on Adult Health and Aging (GERA) Cohort," *Genetics* 200, no. 4 (August 2015):1061–72, doi:10.1534/genetics.115.178624.

2. Rode, L., B. G. Nordestgaard, and S. E. Bojesen, "Peripheral Blood Leukocyte Telomere Length and Mortality Among 64,637 Individuals from the General Population," *Journal of the National Cancer Institute* 107, no. 6 (May 2015): djv074, doi:10.1093/jnci/djv074.

3. Ibid.

4. Lapham et al., "Automated Assay of Telomere Length Measurement and Informatics for 100,000 Subjects in the Genetic Epidemiology Research on Adult Health and Aging (GERA) Cohort." (See #1 above.)

5. Willeit, P., et al., "Leucocyte Telomere Length and Risk of Type 2 Diabetes Mellitus: New Prospective Cohort Study and Literature-Based Meta-analysis," *PLOS ONE* 9, no. 11 (2014): e112483, doi:10.1371/journal.pone.0112483; D'Mello, M. J., et al., "Association Between Shortened Leukocyte Telomere Length and Cardiometabolic Outcomes: Systematic Review and Meta-analysis," *Circulation: Cardiovascular Genetics* 8, no. 1 (February 2015): 82–90, doi:10.1161/CIRCGENET ICS.113.000485; Haycock, P. C., et al., "Leucocyte Telomere Length and Risk of Cardiovascular Disease: Systematic Review and Meta-Analysis," *BMJ* 349 (2014): g4227, doi:10.1136/bmj.g4227; Zhang, C., et al., "The Association Between Telomere Length and Cancer Prognosis: Evidence from a Meta-Analysis," *PLOS ONE* 10, no. 7 (2015): e0133174, doi:10.1371/journal.pone.0133174; and Adnot, S., et al., "Telomere Dysfunction and Cell Senescence in Chronic Lung Diseases: Therapeutic

Potential," *Pharmacology & Therapeutics* 153 (September 2015): 125–34, doi:10.1016/j.pharmthera.2015.06.007.

6. Njajou, O. T., et al., "Association Between Telomere Length, Specific Causes of Death, and Years of Healthy Life in Health, Aging, and Body Composition, a Population-Based Cohort Study," *Journals of Gerontology, Series A: Biological Sciences and Medical Sciences* 64, no. 8 (August 2009): 860–64, doi:10.1093/gerona/glp061.

Chapter Three: Telomerase, the Enzyme That Replenishes Telomeres

1. Vulliamy, T., A. Marrone, F. Goldman, A. Dearlove, M. Bessler, P. J. Mason, and I. Dokal. "The RNA Component of Telomerase Is Mutated in Autosomal Dominant Dyskeratosis Congenita." *Nature* 413, no. 6854 (September 27, 2001): 432–35, doi:10.1038/35096585.

2. Epel, Elissa S., Elizabeth H. Blackburn, Jue Lin, Firdaus S. Dhabhar, Nancy E. Adler, Jason D. Morrow, and Richard M. Cawthon, "Accelerated Telomere Shortening in Response to Life Stress," *Proceedings of the National Academy of Sciences of the United States of America* 101, no. 49 (December 7, 2004): 17312–315, doi:10.1073/pnas.0407162101.

Chapter Four: Unraveling: How Stress Gets into Your Cells

1. Evercare by United Healthcare and the National Alliance for Caregiving, "Evercare Survey of the Economic Downtown and Its Impact on Family Caregiving" (March 2009), 1.

2. Epel, E. S., et al., "Cell Aging in Relation to Stress Arousal and Cardiovascular Disease Risk Factors," *Psychoneuroendocrinology* 31, no. 3 (April 2006): 277–87, doi:10.1016/j.psyneuen.2005.08.011.

3. Gotlib, I. H., et al., "Telomere Length and Cortisol Reactivity in Children of Depressed Mothers," *Molecular Psychiatry* 20, no. 5 (May 2015): 615–20, doi:10.1038/mp.2014.119.

4. Oliveira, B. S., et al., "Systematic Review of the Association between Chronic Social Stress and Telomere Length: A Life Course Perspective," *Ageing Research Reviews* 26 (March 2016): 37–52, doi:10.1016/j.arr.2015.12.006; and Price, L. H., et al., "Telomeres and Early-Life Stress: An Overview." *Biological Psychiatry* 73, no. 1 (January 2013): 15–23, doi:10.1016/j.biopsych.2012.06.025.

5. Mathur, M. B., et al., "Perceived Stress and Telomere Length: A Systematic Review, Meta-analysis, and Methodologic Considerations for Advancing the Field," *Brain, Behavior, and Immunity* 54 (May 2016): 158–69, doi:10.1016/j.bbi.2016.02.002.

6. O'Donovan, A. J., et al., "Stress Appraisals and Cellular Aging: A Key Role for Anticipatory Threat in the Relationship Between Psychological Stress and Telomere Length," *Brain, Behavior, and Immunity* 26, no. 4 (May 2012): 573–79, doi:10.1016/j.bbi.2012.01.007.

7. Ibid.

8. Jefferson, A. L., et al., "Cardiac Index Is Associated with Brain Aging: The Framingham Heart Study," *Circulation* 122, no. 7 (August 17, 2010): 690–97, doi:10.1161/CIRCULATIONAHA.109.905091; and Jefferson, A. L., et al., "Low Cardiac Index Is Associated with Incident Dementia and Alzheimer Disease: The Framingham Heart Study," *Circulation* 131, no. 15 (April 14, 2015): 1333–39, doi:10.1161/CIRCULATIONAHA.114.012438.

9. Sarkar, M., D. Fletcher, D. J. Brown, "What doesn't kill me...": Adversity-Related Experiences Are Vital in the Development of Superior Olympic Performance," *Journal of Science in Medicine and Sport* 18, no. 4 (July 2015): 475–79. doi:10.1016/j.jsams.2014.06.010.

10. Epel, E., et al., "Can Meditation Slow Rate of Cellular Aging? Cognitive Stress, Mindfulness, and Telomeres," *Annals of the New York Academy of Sciences* 1172 (August 2009): 34–53, doi:10.1111/j.1749-6632.2009.04414.x.

11. McLaughlin, K. A., M. A. Sheridan, S. Alves, and W. B. Mendes, "Child Maltreatment and Autonomic Nervous System Reactivity: Identifying Dysregulated Stress Reactivity Patterns by Using the Biopsychosocial Model of Challenge and Threat," *Psychosomatic Medicine* 76, no. 7 (September 2014): 538–46, doi:10.1097/PSY.0000000000000098.

12. O'Donovan et al., "Stress Appraisals and Cellular Aging: A Key Role for Anticipatory Threat in the Relationship Between Psychological Stress and Telomere Length." (See #6 above.)

13. Barrett, L., *How Emotions Are Made* (New York: Houghton Mifflin Harcourt, in press).

14. Ibid.

15. Jamieson, J. P., W. B. Mendes, E. Blackstock, and T. Schmader, "Turning the Knots in Your Stomach into Bows: Reappraising Arousal Improves Performance on the GRE," *Journal of Experimental Social Psychology* 46, no. 1 (January 2010): 208–12.

16. Beltzer, M. L, M. K. Nock, B. J. Peters, and J. P. Jamieson, "Rethinking Butterflies: The Affective, Physiological, and Performance Effects of Reappraising Arousal During Social Evaluation," *Emotion* 14, no. 4 (August 2014): 761–68, doi:10.1037/a0036326.

17. Waugh, C. E., S. Panage, W. B. Mendes, and I. H. Gotlib, "Cardiovascular and Affective Recovery from Anticipatory Threat," *Biological*

Psychology 84, no. 2 (May 2010): 169–175, doi:10.1016/j.biopsycho .2010.01.010; and Lutz, A., et al., "Altered Anterior Insula Activation During Anticipation and Experience of Painful Stimuli in Expert Meditators," *NeuroImage* 64 (January 1, 2013): 538–46, doi:10.1016/ j.neuroimage.2012.09.030.

18. Herborn, K.A., et al., "Stress Exposure in Early Post-Natal Life Reduces Telomere Length: An Experimental Demonstration in a Long-Lived Seabird," *Proceedings of the Royal Society B: Biological Sciences* 281, no. 1782 (March 19, 2014): 20133151, doi:10.1098/rspb.2013.3151.

19. Aydinonat, D., et al., "Social Isolation Shortens Telomeres in African Grey Parrots (*Psittacus erithacus erithacus*)," *PLOS ONE* 9, no. 4 (2014): e93839, doi:10.1371/journal.pone.0093839.

20. Gouin, J. P., L. Hantsoo, and J. K. Kiecolt-Glaser, "Immune Dysregulation and Chronic Stress Among Older Adults: A Review," *Neuroimmunomodulation* 15, nos. 4–6 (2008): 251–59, doi:10.1159/000156468.

21. Cao, W., et al., "Premature Aging of T-Cells Is Associated with Faster HIV-1 Disease Progression," *Journal of Acquired Immune Deficiency Syndromes (1999)* 50, no. 2 (February 1, 2009): 137–47, doi:10.1097 /QAI.0b013e3181926c28.

22. Cohen, S., et al., "Association Between Telomere Length and Experimentally Induced Upper Respiratory Viral Infection in Healthy Adults," *JAMA* 309, no. 7 (February 20, 2013): 699–705, doi:10.1001/jama.2013.613.

23. Choi, J., S. R. Fauce, and R. B. Effros, "Reduced Telomerase Activity in Human T Lymphocytes Exposed to Cortisol," *Brain, Behavior, and Immunity* 22, no. 4 (May 2008): 600–605, doi:10.1016/j.bbi.2007.12.004.

24. Cohen, G. L., and D. K. Sherman, "The Psychology of Change: Self-Affirmation and Social Psychological Intervention," *Annual Review of Psychology* 65 (2014): 333–71, doi:10.1146/annurev-psych-010213- 115137.

25. Miyake, A., et al., "Reducing the Gender Achievement Gap in College Science: A Classroom Study of Values Affirmation," *Science* 330, no. 6008 (November 26, 2010): 1234–37, doi:10.1126/science.1195996.

26. Dutcher, J. M., et al., "Self-Affirmation Activates the Ventral Striatum: A Possible Reward-Related Mechanism for Self-Affirmation," *Psychological Science* 27, no. 4 (April 2016): 455–66, doi:10.1177/ 0956797615625989.

27. Kross, E., et al., "Self-Talk as a Regulatory Mechanism: How You Do It Matters," *Journal of Personality and Social Psychology* 106, no. 2 (February 2014): 304–24, doi:10.1037/a0035173; and Bruehlman-Senecal, E., and O. Ayduk, "This Too Shall Pass: Temporal Distance and the Regulation

of Emotional Distress," *Journal of Personality and Social Psychology* 108, no. 2 (February 2015): 356–75, doi:10.1037/a0038324.

28. Lebois, L. A. M., et al., "A Shift in Perspective: Decentering Through Mindful Attention to Imagined Stressful Events," *Neuropsychologia* 75 (August 2015): 505–24, doi:10.1016/j.neuropsychologia.2015.05.030.

29. Kross, E., et al., " 'Asking Why' from a Distance: Its Cognitive and Emotional Consequences for People with Major Depressive Disorder," *Journal of Abnormal Psychology* 121, no. 3 (August 2012): 559–69, doi:10.1037/a0028808.

Chapter Five: Mind Your Telomeres: Negative Thinking, Resilient Thinking

1. Meyer Friedman and Ray H. Roseman, *Type A Behavior and Your Heart* (New York: Knopf, 1974).

2. Chida, Y., and A. Steptoe, "The Association of Anger and Hostility with Future Coronary Heart Disease: A Meta-analytic Review of Prospective Evidence," *Journal of the American College of Cardiology* 53, no. 11 (March 17, 2009): 936–46, doi:10.1016/j.jacc.2008.11.044.

3. Miller, T. Q, et al., "A Meta-analytic Review of Research on Hostility and Physical Health," *Psychological Bulletin* 119, no. 2 (March 1996): 322–48.

4. Brydon, L., et al., "Hostility and Cellular Aging in Men from the Whitehall II Cohort," *Biological Psychiatry* 71, no. 9 (May 2012): 767–73, doi:10.1016/j.biopsych.2011.08.020.

5. Zalli, A., et al., "Shorter Telomeres with High Telomerase Activity Are Associated with Raised Allostatic Load and Impoverished Psychosocial Resources," *Proceedings of the National Academy of Sciences of the United States of America* 111, no. 12 (March 25, 2014): 4519–24, doi:10.1073/pnas.1322145111.

6. Low, C. A., R. C. Thurston, and K. A. Matthews, "Psychosocial Factors in the Development of Heart Disease in Women: Current Research and Future Directions," *Psychosomatic Medicine* 72, no. 9 (November 2010): 842–54, doi:10.1097/PSY.0b013e3181f6934f.

7. O'Donovan, A., et al., "Pessimism Correlates with Leukocyte Telomere Shortness and Elevated Interleukin-6 in Post-menopausal Women," *Brain, Behavior, and Immunity* 23, no. 4 (May 2009):446–49, doi:10.1016/j.bbi.2008.11.006.

8. Ikeda, A., et al., "Pessimistic Orientation in Relation to Telomere Length in Older Men: The VA Normative Aging Study," *Psychoneuroendocrinology* 42 (April 2014): 68–76, doi:10.1016/j.psyneuen.2014.01.001;

and Schutte, N. S., K. A. Suresh, and J. R. McFarlane, "The Relationship Between Optimism and Longer Telomeres," 2016, under review.

9. Killingsworth, M. A., and D. T. Gilbert, "A Wandering Mind Is an Unhappy Mind," *Science* 330, no. 6006 (November 12, 2010): 932, doi:10.1126/science.1192439.

10. Epel, E. S., et al., "Wandering Minds and Aging Cells," *Clinical Psychological Science* 1, no. 1 (January 2013): 75–83.

11. Kabat-Zinn, J., *Wherever You Go, There You Are: Mindfulness Meditation in Everyday Life* (New York: Hyperion, 1995), p. 15.

12. Engert, V., J. Smallwood, and T. Singer, "Mind Your Thoughts: Associations Between Self-Generated Thoughts and Stress-Induced and Baseline Levels of Cortisol and Alpha-Amylase," *Biological Psychology* 103 (December 2014): 283–91, doi:10.1016/j.biopsycho.2014.10.004.

13. Nolen-Hoeksema, S., "The Role of Rumination in Depressive Disorders and Mixed Anxiety/Depressive Symptoms," *Journal of Abnormal Psychology* 109, no. 3 (August 2000): 504–11.

14. Lea Winerman, "Suppressing the 'White Bears,'" *Monitor on Psychology* 42, no. 9 (October 2011): 44.

15. Alda, M., et al., "Zen Meditation, Length of Telomeres, and the Role of Experiential Avoidance and Compassion," *Mindfulness* 7, no. 3 (June 2016): 651–59.

16. Querstret, D., and M. Cropley, "Assessing Treatments Used to Reduce Rumination and/or Worry: A Systematic Review," *Clinical Psychology Review* 33, no. 8 (December 2013): 996–1009, doi:10.1016/j.cpr.2013.08.004.

17. Wallace, B. Alan, *The Attention Revolution: Unlocking the Power of the Focused Mind* (Boston: Wisdom, 2006).

18. Saron, Clifford, "Training the Mind: The Shamatha Project," in *The Healing Power of Meditation: Leading Experts on Buddhism, Psychology, and Medicine Explore the Health Benefits of Contemplative Practice*, ed. Andy Fraser (Boston: Shambhala, 2013), 45–65.

19. Sahdra, B. K., et al., "Enhanced Response Inhibition During Intensive Meditation Training Predicts Improvements in Self-Reported Adaptive Socioemotional Functioning," *Emotion* 11, no. 2 (April 2011): 299–312, doi:10.1037/a0022764.

20. Schaefer, S. M., et al., "Purpose in Life Predicts Better Emotional Recovery from Negative Stimuli," *PLOS ONE* 8, no. 11 (2013): e80329, doi:10.1371/journal.pone.0080329.

21. Kim, E. S., et al., "Purpose in Life and Reduced Incidence of Stroke in Older Adults: The Health and Retirement Study," *Journal of Psychosomatic*

Research 74, no. 5 (May 2013): 427–32, doi:10.1016/j.jpsychores.2013 .01.013.

22. Boylan, J.M., and C. D. Ryff, "Psychological Wellbeing and Metabolic Syndrome: Findings from the Midlife in the United States National Sample," *Psychosomatic Medicine* 77, no. 5 (June 2015): 548–58, doi:10.1097 /PSY.0000000000000192.

23. Kim, E. S., V. J. Strecher, and C. D. Ryff, "Purpose in Life and Use of Preventive Health Care Services," *Proceedings of the National Academy of Sciences of the United States of America* 111, no. 46 (November 18, 2014): 16331–36, doi:10.1073/pnas.1414826111.

24. Jacobs, T.L., et al., "Intensive Meditation Training, Immune Cell Telomerase Activity, and Psychological Mediators," *Psychoneuroendocrinology* 36, no. 5 (June 2011): 664–81, doi:10.1016/j.psyneuen.2010.09.010.

25. Varma, V. R., et al., "Experience Corps Baltimore: Exploring the Stressors and Rewards of High-Intensity Civic Engagement," *Gerontologist* 55, no. 6 (December 2015): 1038–49, doi:10.1093/geront/gnu011.

26. Gruenewald, T. L., et al., "The Baltimore Experience Corps Trial: Enhancing Generativity via Intergenerational Activity Engagement in Later Life," *Journals of Gerontology, Series B: Psychological Sciences and Social Sciences*, February 25, 2015, doi:10.1093/geronb/gbv005.

27. Carlson, M. C., et al., "Impact of the Baltimore Experience Corps Trial on Cortical and Hippocampal Volumes," *Alzheimer's & Dementia: The Journal of the Alzheimer's Association* 11, no. 11 (November 2015): 1340–48, doi:10.1016/j.jalz.2014.12.005.

28. Sadahiro, R., et al., "Relationship Between Leukocyte Telomere Length and Personality Traits in Healthy Subjects," *European Psychiatry: The Journal of the Association of European Psychiatrists* 30, no. 2 (February 2015): 291–95, doi:10.1016/j.eurpsy.2014.03.003.

29. Edmonds, G. W., H. C. Côté, and S. E. Hampson, "Childhood Conscientiousness and Leukocyte Telomere Length 40 Years Later in Adult Women—Preliminary Findings of a Prospective Association," *PLOS ONE* 10, no. 7 (2015): e0134077, doi:10.1371/journal.pone.0134077.

30. Friedman, H. S., and M. L. Kern, "Personality, Wellbeing, and Health," *Annual Review of Psychology* 65 (2014): 719–42.

31. Costa, D. de S., et al., "Telomere Length Is Highly Inherited and Associated with Hyperactivity-Impulsivity in Children with Attention Deficit/ Hyperactivity Disorder," *Frontiers in Molecular Neuroscience* 8 (2015): 28, doi:10.3389/fnmol.2015.00028; and Yim, O. S., et al., "Delay Discounting, Genetic Sensitivity, and Leukocyte Telomere Length," *Proceedings of*

the *National Academy of Sciences of the United States of America* 113, no. 10 (March 8, 2016): 2780–85, doi:10.1073/pnas.1514351113.

32. Martin, L.R., H. S. Friedman, and J. E. Schwartz, "Personality and Mortality Risk Across the Life Span: The Importance of Conscientiousness as a Biopsychosocial Attribute," *Health Psychology* 26, no. 4 (July 2007): 428–36; and Costa, P. T., Jr., et al., "Personality Facets and All-Cause Mortality Among Medicare Patients Aged 66 to 102 Years: A Follow-On Study of Weiss and Costa (2005)," *Psychosomatic Medicine* 76, no. 5 (June 2014): 370–78, doi:10.1097/PSY.0000000000000070.

33. Shanahan, M. J., et al., "Conscientiousness, Health, and Aging: The Life Course of Personality Model," *Developmental Psychology* 50, no. 5 (May 2014): 1407–25, doi:10.1037/a0031130.

34. Raes, F., E. Pommier, K. D. Neff, and D. Van Gucht, "Construction and Factorial Validation of a Short Form of the Self-Compassion Scale," *Clinical Psychology & Psychotherapy* 18, no. 3 (May–June 2011): 250–55, doi:10.1002/cpp.702.

35. Breines, J. G., et al., "Self-Compassionate Young Adults Show Lower Salivary Alpha-Amylase Responses to Repeated Psychosocial Stress," *Self Identity* 14, no. 4 (October 1, 2015): 390–402.

36. Finlay-Jones, A. L., C. S. Rees, and R. T. Kane, "Self-Compassion, Emotion Regulation and Stress Among Australian Psychologists: Testing an Emotion Regulation Model of Self-Compassion Using Structural Equation Modeling," *PLOS ONE* 10, no. 7 (2015): e0133481, doi:10.1371/journal.pone.0133481.

37. Alda et al., "Zen Meditation, Length of Telomeres, and the Role of Experiential Avoidance and Compassion." (See #15 above.)

38. Hoge, E. A., et al., "Loving-Kindness Meditation Practice Associated with Longer Telomeres in Women," *Brain, Behavior, and Immunity* 32 (August 2013): 159–63, doi:10.1016/j.bbi.2013.04.005.

39. Smeets, E., K. Neff, H. Alberts, and M. Peters, "Meeting Suffering with Kindness: Effects of a Brief Self-Compassion Intervention for Female College Students," *Journal of Clinical Psychology* 70, no. 9 (September 2014): 794–807, doi:10.1002/jclp.22076; and Neff, K. D., and C. K. Germer, "A Pilot Study and Randomized Controlled Trial of the Mindful Self-Compassion Program," *Journal Of Clinical Psychology* 69, no. 1 (January 2013): 28–44, doi:10.1002/jclp.21923.

40. This exercise is adapted from Dr. Neff's website: http://self-compassion .org/exercise-2-self-compassion-break/. For more information on developing self-compassion, see K. Neff, *Self-Compassion: The Proven Power of Being Kind to Yourself* (New York: HarperCollins, 2011).

41. Valenzuela, M., and P. Sachdev, "Can cognitive exercise prevent the onset of dementia? Systematic review of randomized clinical trials with longitudinal follow-up." *Am J Geriatr Psychiatry*, 2009. 17(3): p. 179–87.

Assessment: How Does Your Personality Influence Your Stress Responses?

1. Scheier, M. F., C. S. Carver, and M. W. Bridges, "Distinguishing Optimism from Neuroticism (and Trait Anxiety, Self-Mastery, and Self-Esteem): A Reevaluation of the Life Orientation Test," *Journal of Personality and Social Psychology* 67, no. 6 (December 1994): 1063–78.

2. Marshall, Grant N., et al. "Distinguishing Optimism from Pessimism: Relations to Fundamental Dimensions of Mood and Personality," *Journal of Personality and Social Psychology* 62.6 (1992): 1067.

3. O'Donovan et al., "Pessimism Correlates with Leukocyte Telomere Shortness and Elevated Interleukin-6 in Post-Menopausal Women" (see #7 above); and Ikeda et al., "Pessimistic Orientation in Relation to Telomere Length in Older Men: The VA Normative Aging Study" (see #8 above).

4. Glaesmer, H., et al., "Psychometric Properties and Population-Based Norms of the Life Orientation Test Revised (LOT-R)," *British Journal of Health Psychology* 17, no. 2 (May 2012): 432–45, doi:10.1111/j.2044-8287.2011.02046.x.

5. Eckhardt, Christopher, Bradley Norlander, and Jerry Deffenbacher, "The Assessment of Anger and Hostility: A Critical Review," *Aggression and Violent Behavior* 9, no. 1 (January 2004): 17–43, doi:10.1016/S1359-1789(02)00116-7.

6. Brydon et al., "Hostility and Cellular Aging in Men from the Whitehall II Cohort." (See #4 above.)

7. Trapnell, P. D., and J. D. Campbell, "Private Self-Consciousness and the Five-Factor Model of Personality: Distinguishing Rumination from Reflection," *Journal of Personality and Social Psychology* 76, no. 2 (February 1999) 284–304.

8. Ibid; and Trapnell, P.D., "Rumination-Reflection Questionnaire (RRQ) Shortforms," unpublished data, University of British Columbia (1997).

9. Ibid.

10. John, O. P., E. M. Donahue, and R. L. Kentle, *The Big Five Inventory—Versions 4a and 54* (Berkeley: University of California, Berkeley, Institute of Personality and Social Research, 1991). We thank Dr. Oliver John of UC Berkeley for permission to use this scale. John, O. P., and S.

Srivastava, "The Big-Five Trait Taxonomy: History, Measurement, and Theoretical Perspectives," in *Handbook of Personality: Theory and Research,* ed. L. A. Pervin and O. P. John, 2nd ed. (New York: Guilford Press, 1999): 102–38.

11. Sadahiro, R., et al., "Relationship Between Leukocyte Telomere Length and Personality Traits in Healthy Subjects," *European Psychiatry* 30, no. 2 (February 2015): 291–95, doi:10.1016/j.eurpsy.2014.03.003, pmid: 24768472.

12. Srivastava, S., et al., "Development of Personality in Early and Middle Adulthood: Set Like Plaster or Persistent Change?" *Journal of Personality and Social Psychology* 84, no. 5 (May 2003): 1041–53, doi:10.1037/0022-3514.84.5.1041.

13. Ryff, C. D., and C. L. Keyes, "The Structure of Psychological Wellbeing Revisited," *Journal of Personality and Social Psychology* 69, no. 4 (October 1995): 719–27.

14. Scheier, M. F., et al., "The Life Engagement Test: Assessing Purpose in Life," *Journal of Behavioral Medicine* 29, no. 3 (June 2006): 291–98, doi:10.1007/s10865-005-9044-1.

15. Pearson, E. L., et al., "Normative Data and Longitudinal Invariance of the Life Engagement Test (LET) in a Community Sample of Older Adults," *Quality of Life Research* 22, no. 2 (March 2013): 327–31, doi:10.1007/s11136-012-0146-2.

Chapter Six: When Blue Turns to Gray: Depression and Anxiety

1. Whiteford, H. A., et al., "Global Burden of Disease Attributable to Mental and Substance Use Disorders: Findings from the Global Burden of Disease Study 2010," *Lancet* 382, no. 9904 (November 9, 2013): 1575–86, doi:10.1016/S0140-6736(13)61611-6.

2. Verhoeven, J. E., et al., "Anxiety Disorders and Accelerated Cellular Ageing," *British Journal of Psychiatry* 206, no. 5 (May 2015): 371–78.

3. Cai, N., et al., "Molecular Signatures of Major Depression," *Current Biology* 25, no. 9 (May 4, 2015): 1146–56, doi:10.1016/j.cub.2015.03.008.

4. Verhoeven, J. E., et al., "Major Depressive Disorder and Accelerated Cellular Aging: Results from a Large Psychiatric Cohort Study," *Molecular Psychiatry* 19, no. 8 (August 2014): 895–901, doi:10.1038/mp.2013.151.

5. Mamdani, F., et al., "Variable Telomere Length Across Post-Mortem Human Brain Regions and Specific Reduction in the Hippocampus of Major Depressive Disorder," *Translational Psychiatry* 5 (September 15, 2015): e636, doi:10.1038/tp.2015.134.

6. Zhou, Q. G., et al., "Hippocampal Telomerase Is Involved in the Modulation of Depressive Behaviors," *Journal of Neuroscience* 31, no. 34 (August 24, 2011): 12258–69, doi:10.1523/JNEUROSCI.0805-11.2011.

7. Wolkowitz, O. M., et al., "PBMC Telomerase Activity, but Not Leukocyte Telomere Length, Correlates with Hippocampal Volume in Major Depression," *Psychiatry Research* 232, no. 1 (April 30, 2015): 58–64, doi:10.1016/j.pscychresns.2015.01.007.

8. Darrow, S. M., et al., "The Association between Psychiatric Disorders and Telomere Length: A Meta-analysis Involving 14,827 Persons," *Psychosomatic Medicine* 78, no. 7 (September 2016): 776–87, doi:10.1097/PSY.0000000000000356.

9. Cai et al., "Molecular Signatures of Major Depression." (See #3 above.)

10. Verhoeven, J. E., et al., "The Association of Early and Recent Psychosocial Life Stress with Leukocyte Telomere Length," *Psychosomatic Medicine* 77, no. 8 (October 2015): 882–91, doi:10.1097/PSY.0000000000000226.

11. Verhoeven, J. E., et al., "Major Depressive Disorder and Accelerated Cellular Aging: Results from a Large Psychiatric Cohort Study," *Molecular Psychiatry* 19, no. 8 (August 2014): 895–901, doi:10.1038/mp.2013.151.

12. Ibid.

13. Cai et al., "Molecular Signatures of Major Depression." (See #3 above.)

14. Eisendrath, S. J., et al., "A Preliminary Study: Efficacy of Mindfulness-Based Cognitive Therapy Versus Sertraline as First-Line Treatments for Major Depressive Disorder," *Mindfulness* 6, no. 3 (June 1, 2015): 475–82, doi:10.1007/s12671-014-0280-8; and Kuyken, W., et al., "The Effectiveness and Cost-Effectiveness of Mindfulness-Based Cognitive Therapy Compared with Maintenance Antidepressant Treatment in the Prevention of Depressive Relapse/Recurrence: Results of a Randomised Controlled Trial (the PREVENT Study)," *Health Technology Assessment* 19, no. 73 (September 2015): 1–124, doi:10.3310/hta19730.

15. Teasdale, J. D., et al., "Prevention of Relapse/Recurrence in Major Depression by Mindfulness-Based Cognitive Therapy," *Journal of Consulting and Clinical Psychology* 68, no. 4 (August 2000): 615–23.

16. Teasdale, J., M. Williams, and Z. Segal, *The Mindful Way Workbook: An 8-Week Program to Free Yourself from Depression and Emotional Distress* (New York: Guilford Press, 2014).

17. Wolfson, W., and Epel, E. (2006), "Stress, Post-traumatic Growth, and Leukocyte Aging," poster presentation at the American Psychosomatic Society 64th Annual Meeting, Denver, Colorado, Abstract 1476.

18. Segal, Z., J. M. G. Williams, and J. Teasdale, *Mindfulness-Based Cognitive Therapy for Depression*, 2nd ed. (New York: Guilford Press, 2013), pp. 74–75.

(The three-minute breathing space is part of the MBCT program. Our breathing break is a modified version).

19. Bai, Z., et al., "Investigating the Effect of Transcendental Meditation on Blood Pressure: A Systematic Review and Meta-analysis," *Journal of Human Hypertension* 29, no. 11 (November 2015): 653–62. doi:10.1038/jhh.2015.6; and Cernes, R., and R. Zimlichman, "RESPeRATE: The Role of Paced Breathing in Hypertension Treatment," *Journal of the American Society of Hypertension* 9, no. 1 (January 2015): 38–47, doi:10.1016/j.jash.2014.10.002.

Master Tips for Renewal: Stress-Reducing Techniques Shown to Boost Telomere Maintenance

1. Morgan, N., M. R. Irwin, M. Chung, and C. Wang, "The Effects of Mind-Body Therapies on the Immune System: Meta-analysis," *PLOS ONE* 9, no. 7 (2014): e100903, doi:10.1371/journal.pone.0100903.

2. Conklin, Q., et al., "Telomere Lengthening After Three Weeks of an Intensive Insight Meditation Retreat," *Psychoneuroendocrinology* 61 (November 2015): 26–27, doi:10.1016/j.psyneuen.2015.07.462.

3. Epel, E., et al. "Meditation and Vacation Effects Impact Disease-Associated Molecular Phenotypes," *Translational Psychiatry* (August 2016): 6, e880, doi: 10.1038/tp.2016.164.

4. Kabat-Zinn, J., *Full Catastrophe Living: Using the Wisdom of Your Body and Mind to Face Stress, Pain, and Illness*, rev. ed. (New York: Bantam Books, 2013).

5. Lengacher, C. A., et al., "Influence of Mindfulness-Based Stress Reduction (MBSR) on Telomerase Activity in Women with Breast Cancer (BC)," *Biological Research for Nursing* 16, no. 4 (October 2014): 438–47, doi:10.1177/1099800413519495.

6. Carlson, L. E., et al., "Mindfulness-Based Cancer Recovery and Supportive-Expressive Therapy Maintain Telomere Length Relative to Controls in Distressed Breast Cancer Survivors," *Cancer* 121, no. 3 (February 1, 2015): 476–84, doi:10.1002/cncr.29063.

7. Black, D. S., et al., "Yogic Meditation Reverses NF-κB- and IRF-Related Transcriptome Dynamics in Leukocytes of Family Dementia Caregivers in a Randomized Controlled Trial," *Psychoneuroendocrinology* 38, no. 3 (March 2013): 348–55, doi:10.1016/j.psyneuen.2012.06.011.

8. Lavretsky, H., et al.,"A Pilot Study of Yogic Meditation for Family Dementia Caregivers with Depressive Symptoms: Effects on Mental Health, Cognition, and Telomerase Activity," *International Journal of Geriatric Psychiatry* 28, no. 1 (January 2013): 57–65, doi:10.1002/gps.3790.

9. Desveaux, L., A. Lee, R. Goldstein, and D. Brooks, "Yoga in the Management of Chronic Disease: A Systematic Review and

Meta-analysis," *Medical Care* 53, no. 7 (July 2015): 653–61, doi:10.1097/MLR.0000000000000372.

10. Hartley, L., et al., "Yoga for the Primary Prevention of Cardiovascular Disease," *Cochrane Database of Systematic Reviews* 5 (May 13, 2014): CD010072, doi:10.1002/14651858.CD010072.pub2.

11. Lu, Y. H., B. Rosner, G. Chang, and L. M. Fishman, "Twelve-Minute Daily Yoga Regimen Reverses Osteoporotic Bone Loss," *Topics in Geriatric Rehabilitation* 32, no. 2 (April 2016): 81–87.

12. Liu, X., et al., "A Systematic Review and Meta-analysis of the Effects of Qigong and Tai Chi for Depressive Symptoms," *Complementary Therapies in Medicine* 23, no. 4 (August 2015): 516–34, doi:10.1016/j.ctim.2015.05.001.

13. Freire, M. D., and C. Alves, "Therapeutic Chinese Exercises (Qigong) in the Treatment of Type 2 Diabetes Mellitus: A Systematic Review," *Diabetes & Metabolic Syndrome: Clinical Research & Reviews* 7, no. 1 (March 2013): 56–59, doi:10.1016/j.dsx.2013.02.009.

14. Ho, R. T. H., et al., "A Randomized Controlled Trial of Qigong Exercise on Fatigue Symptoms, Functioning, and Telomerase Activity in Persons with Chronic Fatigue or Chronic Fatigue Syndrome," *Annals of Behavioral Medicine* 44, no. 2 (October 2012): 160–70, doi:10.1007/s12160-012-9381-6.

15. Ornish D., et al., "Effect of Comprehensive Lifestyle Changes on Telomerase Activity and Telomere Length in Men with Biopsy-Proven Low-Risk Prostate Cancer: 5-Year Follow-Up of a Descriptive Pilot Study," *Lancet Oncology* 14, no. 11 (October 2013): 1112–20, doi:10.1016/S1470-2045(13)70366-8.

Assessment: What's Your Telomere Trajectory? Protective and Risky Factors

1. Ahola, K., et al., "Work-Related Exhaustion and Telomere Length: A Population-Based Study," *PLOS ONE* 7, no. 7 (2012): e40186, doi:10.1371/journal.pone.0040186.

2. Damjanovic, A. K., et al., "Accelerated Telomere Erosion Is Associated with a Declining Immune Function of Caregivers of Alzheimer's Disease Patients," *Journal of Immunology* 179, no. 6 (September 15, 2007): 4249–54.

3. Geronimus, A. T., et al., "Race-Ethnicity, Poverty, Urban Stressors, and Telomere Length in a Detroit Community-Based Sample," *Journal of Health and Social Behavior* 56, no. 2 (June 2015): 199–224, doi:10.1177/0022146515582100.

4. Darrow, S. M., et al., "The Association between Psychiatric Disorders and Telomere Length: A Meta-analysis Involving 14,827 Persons,"

Psychosomatic Medicine 78, no. 7 (September 2016): 776–87, doi:10.1097 /PSY.0000000000000356; and Lindqvist et al, "Psychiatric Disorders and Leukocyte Telomere Length: Underlying Mechanisms Linking Mental Illness with Cellular Aging," *Neuroscience & Biobehavioral Reviews* 55 (August 2015): 333–64, doi:10.1016/j.neubiorev.2015.05.007.

5. Mitchell, P. H., et al., "A Short Social Support Measure for Patients Recovering from Myocardial Infarction: The ENRICHD Social Support Inventory," *Journal of Cardiopulmonary Rehabilitation* 23, no. 6 (November–December 2003): 398–403.

6. Zalli, A., et al., "Shorter Telomeres with High Telomerase Activity Are Associated with Raised Allostatic Load and Impoverished Psychosocial Resources," *Proceedings of the National Academy of Sciences of the United States of America* 111, no. 12 (March 25, 2014): 4519–24, doi:10.1073 /pnas.1322145111; and Carroll, J. E., A. V. Diez Roux, A. L. Fitzpatrick, and T. Seeman, "Low Social Support Is Associated with Shorter Leukocyte Telomere Length in Late Life: Multi-Ethnic Study of Atherosclerosis," *Psychosomatic Medicine* 75, no. 2 (February 2013): 171–77, doi:10.1097/PSY.0b013e31828233bf.

7. Carroll et al., "Low Social Support Is Associated with Shorter Leukocyte Telomere Length in Late Life: Multi-ethnic Study of Atherosclerosis." (See #6 above.)

8. Kiernan, M., et al., "The Stanford Leisure-Time Activity Categorical Item (L-Cat): A Single Categorical Item Sensitive to Physical Activity Changes in Overweight/Obese Women," *International Journal of Obesity (2005)* 37, no. 12 (December 2013): 1597–1602, doi:10.1038/ijo.2013.36.

9. Puterman, E., et al., "The Power of Exercise: Buffering the Effect of Chronic Stress on Telomere Length," *PLOS ONE* 5, no. 5 (2010): e10837, doi:10.1371/journal.pone.0010837; and Puterman, E., et al., "Determinants of Telomere Attrition over One Year in Healthy Older Women: Stress and Health Behaviors Matter," *Molecular Psychiatry* 20, no. 4 (April 2015): 529–35, doi:10.1038/mp.2014.70.

10. Werner, C., A. Hecksteden, J. Zundler, M. Boehm, T. Meyer, and U. Laufs. "Differential Effects of Aerobic Endurance, Interval and Strength Endurance Training on Telomerase Activity and Senescence Marker Expression in Circulating Mononuclear Cells." *European Heart Journal* 36 (2015) (Abstract Supplement): P2370. Manuscript in progress.

11. Buysse D. J., et al., "The Pittsburgh Sleep Quality Index: A New Instrument for Psychiatric Practice and Research," *Psychiatry Research* 28, no. 2 (May 1989): 193–213.

12. Prather, A. A., et al., "Tired Telomeres: Poor Global Sleep Quality, Perceived Stress, and Telomere Length in Immune Cell Subsets in Obese Men and Women," *Brain, Behavior, and Immunity* 47 (July 2015): 155–162, doi:10.1016/j.bbi.2014.12.011.

13. Farzaneh-Far, R., et al., "Association of Marine Omega-3 Fatty Acid Levels with Telomeric Aging in Patients with Coronary Heart Disease," *JAMA* 303, no. 3 (January 20, 2010): 250–57, doi:10.1001/jama.2009.2008.

14. Lee, J. Y., et al., "Association Between Dietary Patterns in the Remote Past and Telomere Length," *European Journal of Clinical Nutrition* 69, no. 9 (September 2015): 1048–52, doi:10.1038/ejcn.2015.58.

15. Kiecolt-Glaser, J. K., et al., "Omega-3 Fatty Acids, Oxidative Stress, and Leukocyte Telomere Length: A Randomized Controlled Trial," *Brain, Behavior, and Immunity* 28 (February 2013): 16–24, doi:10.1016/j.bbi.2012.09.004.

16. Lee, "Association between Dietary Patterns in the Remote Past and Telomere Length" (see #14 above); Leung, C. W., et al., "Soda and Cell Aging: Associations Between Sugar-Sweetened Beverage Consumption and Leukocyte Telomere Length in Healthy Adults from the National Health and Nutrition Examination Surveys," *American Journal of Public Health* 104, no. 12 (December 2014): 2425–31, doi:10.2105/AJPH.2014.302151; and Leung, C., et al., "Sugary Beverage and Food Consumption and Leukocyte Telomere Length Maintenance in Pregnant Women," *European Journal of Clinical Nutrition* (June 2016): doi:10.1038/ejcn.2016.v93.

17. Nettleton, J. A., et al., "Dietary Patterns, Food Groups, and Telomere Length in the Multi-ethnic Study of Atherosclerosis (MESA)," *American Journal of Clinical Nutrition* 88, no. 5 (November 2008): 1405–12.

18. Valdes, A. M., et al., "Obesity, Cigarette Smoking, and Telomere Length in Women," *Lancet* 366, no. 9486 (August 20–26, 2005): 662–664; and McGrath, M., et al., "Telomere Length, Cigarette Smoking, and Bladder Cancer Risk in Men and Women," *Cancer Epidemiology, Biomarkers, and Prevention* 16, no. 4 (April 2007): 815–19.

19. Kahl, V. F., et al., "Telomere Measurement in Individuals Occupationally Exposed to Pesticide Mixtures in Tobacco Fields," *Environmental and Molecular Mutagenesis* 57, no. 1 (January 2016): 74–84, doi:10.1002/em.21984.

20. Pavanello, S., et al., "Shorter Telomere Length in Peripheral Blood Lymphocytes of Workers Exposed to Polycyclic Aromatic Hydrocarbons,"

Carcinogenesis 31, no. 2 (February 2010): 216–21, doi:10.1093/carcin/bgp278.

21. Hou, L., et al., "Air Pollution Exposure and Telomere Length in Highly Exposed Subjects in Beijing, China: A Repeated-Measure Study," *Environment International* 48 (November 1, 2012): 71–77, doi:10.1016/j.envint.2012.06.020; and Hoxha, M., et al., "Association between Leukocyte Telomere Shortening and Exposure to Traffic Pollution: A Cross-Sectional Study on Traffic Officers and Indoor Office Workers," *Environmental Health* 8 (September 21, 2009): 41, doi:10.1186/1476-069X-8-41.

22. Wu, Y., et al., "High Lead Exposure Is Associated with Telomere Length Shortening in Chinese Battery Manufacturing Plant Workers," *Occupational and Environmental Medicine* 69, no. 8 (August 2012): 557–63, doi:10.1136/oemed-2011-100478.

23. Pavanello et al., "Shorter Telomere Length in Peripheral Blood Lymphocytes of Workers Exposed to Polycyclic Aromatic Hydrocarbons" (see #20 above); and Bin, P., et al., "Association Between Telomere Length and Occupational Polycyclic Aromatic Hydrocarbons Exposure," *Zhonghua Yu Fang Yi Xue Za Zhi* 44, no. 6 (June 2010): 535–38. (The article is in Chinese.)

Chapter Seven: Training Your Telomeres: How Much Exercise Is Enough?

1. Najarro, K., et al., "Telomere Length as an Indicator of the Robustness of B- and T-Cell Response to Influenza in Older Adults," *Journal of Infectious Diseases* 212, no. 8 (October 15, 2015): 1261–69, doi:10.1093/infdis/jiv202.

2. Simpson, R. J., et al., "Exercise and the Aging Immune System," *Ageing Research Reviews* 11, no. 3 (July 2012): 404–20, doi:10.1016/j.arr.2012.03.003.

3. Cherkas, L. F., et al., "The Association between Physical Activity in Leisure Time and Leukocyte Telomere Length," *Archives of Internal Medicine* 168, no. 2 (January 28, 2008): 154–58, doi:10.1001/archinternmed.2007.39.

4. Loprinzi, P. D., "Leisure-Time Screen-Based Sedentary Behavior and Leukocyte Telomere Length: Implications for a New Leisure-Time Screen-Based Sedentary Behavior Mechanism," *Mayo Clinic Proceedings* 90, no. 6 (June 2015): 786–90, doi:10.1016/j.mayocp.2015.02.018; and Sjögren, P., et al., "Stand Up for Health—Avoiding Sedentary Behaviour Might Lengthen Your Telomeres: Secondary Outcomes from a Physical

Activity RCT in Older People," *British Journal of Sports Medicine* 48, no 19 (October 2014): 1407–09, doi:10.1136/bjsports-2013-093342.

5. Werner, C., et al., "Differential Effects of Aerobic Endurance, Interval and Strength Endurance Training on Telomerase Activity and Senescence Marker Expression in Circulating Mononuclear Cells," *European Heart Journal* 36 (abstract supplement) (August 2015): P2370, http://eur heartj.oxfordjournals.org/content/ehj/36/suppl_1/163.full.pdf.

6. Loprinzi, P. D., J. P. Loenneke, and E. H. Blackburn, "Movement-Based Behaviors and Leukocyte Telomere Length among US Adults," *Medicine and Science in Sports and Exercise* 47, no. 11 (November 2015): 2347–52, doi:10.1249/MSS.0000000000000695.

7. Chilton, W. L., et al., "Acute Exercise Leads to Regulation of Telomere-Associated Genes and MicroRNA Expression in Immune Cells," *PLOS ONE* 9, no. 4 (2014): e92088, doi:10.1371/journal.pone.0092088.

8. Denham, J., et al., "Increased Expression of Telomere-Regulating Genes in Endurance Athletes with Long Leukocyte Telomeres," *Journal of Applied Physiology (1985)* 120, no. 2 (January 15, 2016): 148–58, doi:10.1152/japplphysiol.00587.2015.

9. Rana, K. S., et al., "Plasma Irisin Levels Predict Telomere Length in Healthy Adults," *Age* 36, no. 2 (April 2014): 995–1001, doi:10.1007/s11357-014-9620-9.

10. Mooren, F. C., and K. Krüger, "Exercise, Autophagy, and Apoptosis," *Progress in Molecular Biology and Translational Science* 135 (2015): 407–22, doi:10.1016/bs.pmbts.2015.07.023.

11. Hood, D. A., et al., "Exercise and the Regulation of Mitochondrial Turnover," *Progress in Molecular Biology and Translational Science* 135 (2015): 99–127, doi:10.1016/bs.pmbts.2015.07.007.

12. Loprinzi, P. D., "Cardiorespiratory Capacity and Leukocyte Telomere Length Among Adults in the United States," *American Journal of Epidemiology* 182, no. 3 (August 1, 2015): 198–201, doi:10.1093/aje/kwv056.

13. Krauss, J., et al., "Physical Fitness and Telomere Length in Patients with Coronary Heart Disease: Findings from the Heart and Soul Study," *PLOS ONE* 6, no. 11 (2011): e26983, doi:10.1371/journal.pone.0026983.

14. Denham, J., et al., "Longer Leukocyte Telomeres Are Associated with Ultra-Endurance Exercise Independent of Cardiovascular Risk Factors," *PLOS ONE* 8, no. 7 (2013): e69377, doi:10.1371/journal.pone.0069377.

15. Denham et al., "Increased Expression of Telomere-Regulating Genes in Endurance Athletes with Long Leukocyte Telomeres." (See #8 above.)

16. Laine, M. K., et al., "Effect of Intensive Exercise in Early Adult Life on Telomere Length in Later Life in Men," *Journal of Sports Science and Medicine* 14, no. 2 (June 2015): 239–45.

17. Werner, C., et al., "Physical Exercise Prevents Cellular Senescence in Circulating Leukocytes and in the Vessel Wall," *Circulation* 120, no. 24 (December 15, 2009): 2438–47, doi:10.1161/CIRCULATIONAHA.109.861005.

18. Saßenroth, D., et al., "Sports and Exercise at Different Ages and Leukocyte Telomere Length in Later Life—Data from the Berlin Aging Study II (BASE-II)," *PLOS ONE* 10, no. 12 (2015): e0142131, doi:10.1371/journal.pone.0142131.

19. Collins, M., et al., "Athletes with Exercise-Associated Fatigue Have Abnormally Short Muscle DNA Telomeres," *Medicine and Science in Sports and Exercise* 35, no. 9 (September 2003): 1524–28.

20. Wichers, M., et al., "A Time-Lagged Momentary Assessment Study on Daily Life Physical Activity and Affect," *Health Psychology* 31, no. 2 (March 2012): 135–144, doi:10.1037/a0025688.

21. Von Haaren, B., et al., "Does a 20-Week Aerobic Exercise Training Programme Increase Our Capabilities to Buffer Real-Life Stressors? A Randomized, Controlled Trial Using Ambulatory Assessment," *European Journal of Applied Physiology* 116, no. 2 (February 2016): 383–94, doi:10.1007/s00421-015-3284-8.

22. Puterman, E., et al., "The Power of Exercise: Buffering the Effect of Chronic Stress on Telomere Length," *PLOS ONE* 5, no. 5 (2010): e10837, doi:10.1371/journal.pone.0010837.

23. Puterman, E., et al., "Multisystem Resiliency Moderates the Major Depression–Telomere Length Association: Findings from the Heart and Soul Study," *Brain, Behavior, and Immunity* 33 (October 2013): 65–73, doi:10.1016/j.bbi.2013.05.008.

24. Werner et al., "Differential Effects of Aerobic Endurance, Interval and Strength Endurance Training on Telomerase Activity and Senescence Marker Expression in Circulating Mononuclear Cells." (See #5 above.)

25. Masuki, S., et al., "The Factors Affecting Adherence to a Long-Term Interval Walking Training Program in Middle-Aged and Older People," *Journal of Applied Physiology (1985)* 118, no. 5 (March 1, 2015): 595–603, doi:10.1152/japplphysiol.00819.2014.

26. Loprinzi, "Leisure-Time Screen–Based Sedentary Behavior and Leukocyte Telomere Length." (See #4 above.)

Chapter Eight: Tired Telomeres: From Exhaustion to Restoration

1. "Lack of Sleep Is Affecting Americans, Finds the National Sleep Foundation," National Sleep Foundation, https://sleepfoundation.org /media-center/press-release/lack-sleep-affecting-americans-finds-the -national-sleep-foundation, accessed September 29, 2015.

2. Carroll, J. E., et al., "Insomnia and Telomere Length in Older Adults," *Sleep* 39, no. 3 (March 1, 2016): 559–64, doi:10.5665/sleep.5526.

3. Micic, G., et al., "The Etiology of Delayed Sleep Phase Disorder," *Sleep Medicine Reviews* 27 (June 2016): 29–38, doi:10.1016/j.smrv.2015.06.004.

4. Sachdeva, U. M., and C. B. Thompson, "Diurnal Rhythms of Autophagy: Implications for Cell Biology and Human Disease," *Autophagy* 4, no. 5 (July 2008): 581–89.

5. Gonnissen, H. K. J., T. Hulshof, and M. S. Westerterp-Plantenga, "Chronobiology, Endocrinology, and Energy-and-Food-Reward Homeostasis," *Obesity Reviews* 14, no. 5 (May 2013): 405–16, doi:10.1111 /obr.12019.

6. Van der Helm, E., and M. P. Walker, "Sleep and Emotional Memory Processing," *Journal of Clinical Sleep Medicine* 6, no. 1 (March 2011): 31–43.

7. Meerlo, P., A. Sgoifo, and D. Suchecki, "Restricted and Disrupted Sleep: Effects on Autonomic Function, Neuroendocrine Stress Systems and Stress Responsivity," *Sleep Medicine Reviews* 12, no. 3 (June 2008): 197–210, doi:10.1016/j.smrv.2007.07.007.

8. Walker, M. P., "Sleep, Memory, and Emotion," *Progress in Brain Research* 185 (2010): 49–68, doi:10.1016/B978-0-444-53702-7.00004-X.

9. Lee, K. A., et al., "Telomere Length Is Associated with Sleep Duration but Not Sleep Quality in Adults with Human Immunodeficiency Virus," *Sleep* 37, no. 1 (January 1, 2014): 157–66, doi:10.5665/sleep.3328; and Cribbet, M. R., et al., "Cellular Aging and Restorative Processes: Subjective Sleep Quality and Duration Moderate the Association between Age and Telomere Length in a Sample of Middle-Aged and Older Adults," *Sleep* 37, no. 1 (January 1, 2014): 65–70, doi:10.5665/sleep.3308.

10. Jackowska, M., et. al., "Short Sleep Duration Is Associated with Shorter Telomere Length in Healthy Men: Findings from the Whitehall II Cohort Study," *PLOS ONE* 7, no. 10 (2012): e47292, doi:10.1371/journal .pone.0047292.

11. Cribbet et al., "Cellular Aging and Restorative Processes." (See #9 above.)

12. Ibid.

13. Prather, A. A., et al., "Tired Telomeres: Poor Global Sleep Quality, Perceived Stress, and Telomere Length in Immune Cell Subsets in Obese Men and Women," *Brain, Behavior, and Immunity* 47 (July 2015): 155–62, doi:10.1016/j.bbi.2014.12.011.

14. Chen, W. D., et al., "The Circadian Rhythm Controls Telomeres and Telomerase Activity," *Biochemical and Biophysical Research Communications* 451, no. 3 (August 29, 2014): 408–14, doi:10.1016/j.bbrc.2014.07.138.

15. Ong, J., and D. Sholtes, "A Mindfulness-Based Approach to the Treatment of Insomnia," *Journal of Clinical Psychology* 66, no. 11 (November 2010): 1175–84, doi:10.1002/jclp.20736.

16. Ong, J. C., et al., "A Randomized Controlled Trial of Mindfulness Meditation for Chronic Insomnia," *Sleep* 37, no. 9 (September 1, 2014): 1553–63B, doi:10.5665/sleep.4010.

17. Chang, A. M., D. Aeschbach, J. F. Duffy, and C. A. Czeisler, "Evening Use of Light-Emitting eReaders Negatively Affects Sleep, Circadian Timing, and Next-Morning Alertness," *Proceedings of the National Academy of Sciences of the United States of America* 112, no. 4 (January 2015): 1232–37, doi:10.1073/pnas.1418490112.

18. Dang-Vu, T. T., et al., "Spontaneous Brain Rhythms Predict Sleep Stability in the Face of Noise," *Current Biology* 20, no. 15 (August 10, 2010): R626–27, doi:10.1016/j.cub.2010.06.032.

19. Griefhan, B., P. Bröde, A. Marks, and M. Basner, "Autonomic Arousals Related to Traffic Noise During Sleep," *Sleep* 31, no. 4 (April 2008): 569–77.

20. Savolainen, K., et al., "The History of Sleep Apnea Is Associated with Shorter Leukocyte Telomere Length: The Helsinki Birth Cohort Study," *Sleep Medicine* 15, no. 2 (February 2014): 209–12, doi:10.1016/j.sleep.2013.11.779.

21. Salihu, H. M., et al., "Association Between Maternal Symptoms of Sleep Disordered Breathing and Fetal Telomere Length," *Sleep* 38, no. 4 (April 1, 2015): 559–66, doi:10.5665/sleep.4570.

22. Shin, C., C. H. Yun, D. W. Yoon, and I. Baik, "Association Between Snoring and Leukocyte Telomere Length," *Sleep* 39, no. 4 (April 1, 2016): 767–72, doi:10.5665/sleep.5624.

Chapter Nine: Telomeres Weigh In: A Healthy Metabolism

1. Mundstock, E., et al., "Effect of Obesity on Telomere Length: Systematic Review and Meta-analysis," *Obesity (Silver Spring)* 23, no. 11 (November 2015): 2165–74, doi:10.1002/oby.21183.

2. Bosello, O., M. P. Donataccio, and M. Cuzzolaro, "Obesity or Obesities? Controversies on the Association Between Body Mass Index and

Premature Mortality," *Eating and Weight Disorders* 21, no. 2 (June 2016): 165–74, doi:10.1007/s40519-016-0278-4.

3. Farzaneh-Far, R., et al., "Telomere Length Trajectory and Its Determinants in Persons with Coronary Artery Disease: Longitudinal Findings from the Heart and Soul Study," *PLOS ONE* 5, no. 1 (January 2010): e8612, doi:10.1371/journal.pone.0008612.

4. "IDF Diabetes Atlas, Sixth Edition," *International Diabetes Federation*, http://www.idf.org/atlasmap/atlasmap?indicator=i1&date=2014, accessed September 16, 2015.

5. Farzaneh-Far et al., "Telomere Length Trajectory and Its Determinants in Persons with Coronary Artery Disease." (See #3 above.)

6. Verhulst, S., et al., "A Short Leucocyte Telomere Length Is Associated with Development of Insulin Resistance," *Diabetologia* 59, no. 6 (June 2016): 1258–65, doi:10.1007/s00125-016-3915-6.

7. Zhao, J., et al., "Short Leukocyte Telomere Length Predicts Risk of Diabetes in American Indians: The Strong Heart Family Study," *Diabetes* 63, no. 1 (January 2014): 354–62, doi:10.2337/db13-0744.

8. Willeit, P., et al., "Leucocyte Telomere Length and Risk of Type 2 Diabetes Mellitus: New Prospective Cohort Study and Literature-Based Meta-analysis," *PLOS ONE* 9, no. 11 (2014): e112483, doi:10.1371/journal.pone.0112483.

9. Guo, N., et al., "Short Telomeres Compromise β-Cell Signaling and Survival," *PLOS ONE* 6, no. 3 (2011): e17858, doi:10.1371/journal.pone.0017858.

10. Formichi, C., et al., "Weight Loss Associated with Bariatric Surgery Does Not Restore Short Telomere Length of Severe Obese Patients after 1 Year," *Obesity Surgery* 24, no. 12 (December 2014): 2089–93, doi:10.1007/s11695-014-1300-4.

11. Gardner, J. P., et al., "Rise in Insulin Resistance is Associated with Escalated Telomere Attrition," *Circulation* 111, no. 17 (May 3, 2005): 2171–77.

12. Fothergill, Erin, Juen Guo, Lilian Howard, Jennifer C. Kerns, Nicolas D. Knuth, Robert Brychta, Kong Y. Chen, et al. "Persistent Metabolic Adaptation Six Years after *The Biggest Loser* Competition," *Obesity* (Silver Spring, Md.), May 2, 2016, doi:10.1002/oby.21538.

13. Kim, S., et al., "Obesity and Weight Gain in Adulthood and Telomere Length," *Cancer Epidemiology, Biomarkers & Prevention* 18, no. 3 (March 2009): 816–20, doi:10.1158/1055-9965.EPI-08-0935.

14. Cottone, P., et al., "CRF System Recruitment Mediates Dark Side of Compulsive Eating," *Proceedings of the National Academy of Sciences of*

the United States of America 106, no. 47 (November 2009): 20016–20, doi:0.1073/pnas.0908789106.

15. Tomiyama, A. J., et al., "Low Calorie Dieting Increases Cortisol," *Psychosomatic Medicine* 72, no. 4 (May 2010): 357–64, doi:10.1097 /PSY.0b013e3181d9523c.

16. Kiefer, A., J. Lin, E. Blackburn, and E. Epel, "Dietary Restraint and Telomere Length in Pre- and Post-Menopausal Women," *Psychosomatic Medicine* 70, no. 8 (October 2008): 845–49, doi:10.1097/PSY.0b013 e318187d05e.

17. Hu, F. B., "Resolved: There Is Sufficient Scientific Evidence That Decreasing Sugar-Sweetened Beverage Consumption Will Reduce the Prevalence of Obesity and Obesity-Related Diseases," *Obesity Reviews* 14, no. 8 (August 2013): 606–19, doi:10.1111/obr.12040; and Yang, Q., et al., "Added Sugar Intake and Cardiovascular Diseases Mortality Among U.S. Adults," *JAMA Internal Medicine* 174, no. 4 (April 2014): 516–24, doi:10.1001/jamainternmed.2013.13563.

18. Schulte, E. M., N. M. Avena, and A. N. Gearhardt, "Which Foods May Be Addictive? The Roles of Processing, Fat Content, and Glycemic Load," *PLOS ONE* 10, no. 2 (February 18, 2015): e0117959, doi:10.1371 /journal.pone.0117959.

19. Lustig, R. H., et al., "Isocaloric Fructose Restriction and Metabolic Improvement in Children with Obesity and Metabolic Syndrome," *Obesity* 2 (February 24, 2016): 453–60, doi:10.1002/oby.21371, epub October 26, 2015.

20. Incollingo Belsky, A. C., E. S. Epel, and A. J. Tomiyama, "Clues to Maintaining Calorie Restriction? Psychosocial Profiles of Successful Long-Term Restrictors," *Appetite* 79 (August 2014): 106–12, doi:10.1016 /j.appet.2014.04.006.

21. Wang, C., et al., "Adult-Onset, Short-Term Dietary Restriction Reduces Cell Senescence in Mice," *Aging* 2, no. 9 (September 2010): 555–66.

22. Daubenmier, J., et al., "Changes in Stress, Eating, and Metabolic Factors Are Related to Changes in Telomerase Activity in a Randomized Mindfulness Intervention Pilot Study," *Psychoneuroendocrinology* 37, no. 7 (July 2012): 917–28, doi:10.1016/j.psyneuen.2011.10.008.

23. Mason, A. E., et al., "Effects of a Mindfulness-Based Intervention on Mindful Eating, Sweets Consumption, and Fasting Glucose Levels in Obese Adults: Data from the SHINE Randomized Controlled Trial," *Journal of Behavioral Medicine* 39, no. 2 (April 2016): 201–13, doi:10.1007/s10865-015-9692-8.

24. Kristeller, J., with A. Bowman, *The Joy of Half a Cookie: Using Mindfulness to Lose Weight and End the Struggle with Food* (New York: Perigee, 2015). Also see www.mindfuleatingtraining.com and www.mb-eat.com.

Chapter Ten: Food and Telomeres: Eating for Optimal Cell Health

1. Jurk, D., et al., "Chronic Inflammation Induces Telomere Dysfunction and Accelerates Ageing in Mice," *Nature Communications* 2 (June 24, 2104): 4172, doi:10.1038/ncomms5172.

2. "What You Eat Can Fuel or Cool Inflammation, A Key Driver of Heart Disease, Diabetes, and Other Chronic Conditions," Harvard Medical School, Harvard Health Publications, http://www.health.harvard.edu/family_health_guide/what-you-eat-can-fuel-or-cool-inflammation-a-key-driver-of-heart-disease-diabetes-and-other-chronic-conditions, accessed November 27, 2015.

3. Weischer, M., S. E. Bojesen, and B. G. Nordestgaard, "Telomere Shortening Unrelated to Smoking, Body Weight, Physical Activity, and Alcohol Intake: 4,576 General Population Individuals with Repeat Measurements 10 Years Apart," *PLOS Genetics* 10, no. 3 (March 13, 2014): e1004191, doi:10.1371/journal.pgen.1004191; and Pavanello, S., et al., "Shortened Telomeres in Individuals with Abuse in Alcohol Consumption," *International Journal of Cancer* 129, no. 4 (August 15, 2011): 983–92. doi:10.1002/ijc.25999.

4. Cassidy, A., et al., "Higher Dietary Anthocyanin and Flavonol Intakes Are Associated with Anti-inflammatory Effects in a Population of U.S. Adults," *American Journal of Clinical Nutrition* 102, no. 1 (July 2015): 172–81, doi:10.3945/ajcn.115.108555.

5. Farzaneh-Far, R., et al., "Association of Marine Omega-3 Fatty Acid Levels with Telomeric Aging in Patients with Coronary Heart Disease," *JAMA* 303, no. 3 (January 20, 2010): 250–57, doi:10.1001/jama.2009.2008.

6. Goglin, S., et al., "Leukocyte Telomere Shortening and Mortality in Patients with Stable Coronary Heart Disease from the Heart and Soul Study," *PLOS ONE* (2016), in press.

7. Farzaneh-Far et al., "Association of Marine Omega-3 Fatty Acid Levels with Telomeric Aging in Patients with Coronary Heart Disease." (See #5 above.)

8. Kiecolt-Glaser, J. K., et. al., "Omega-3 Fatty Acids, Oxidative Stress, and Leukocyte Telomere Length: A Randomized Controlled Trial," *Brain, Behavior, and Immunity* 28 (February 2013): 16–24, doi:10.1016/j.bbi.2012.09.004.

9. Glei, D. A., et al., "Shorter Ends, Faster End? Leukocyte Telomere Length and Mortality Among Older Taiwanese," *Journals of Gerontology, Series A: Biological Sciences and Medical Sciences* 70, no. 12 (December 2015): 1490–98, doi:10.1093/gerona/glu191.

10. Debreceni, B., and L. Debreceni, "The Role of Homocysteine-Lowering B-Vitamins in the Primary Prevention of Cardiovascular Disease," *Cardiovascular Therapeutics* 32, no. 3 (June 2014): 130–38, doi:10.1111/1755 -5922.12064.

11. Kawanishi, S., and S. Oikawa, "Mechanism of Telomere Shortening by Oxidative Stress," *Annals of the New York Academy of Sciences* 1019 (June 2004): 278–84.

12. Haendeler, J., et al., "Hydrogen Peroxide Triggers Nuclear Export of Telomerase Reverse Transcriptase via Src Kinase Familiy-Dependent Phosphorylation of Tyrosine 707," *Molecular and Cellular Biology* 23, no. 13 (July 2003): 4598–610.

13. Adelfalk, C., et al., "Accelerated Telomere Shortening in Fanconi Anemia Fibroblasts—a Longitudinal Study," *FEBS Letters* 506, no. 1 (September 28, 2001): 22–26.

14. Xu, Q., et al., "Multivitamin Use and Telomere Length in Women," *American Journal of Clinical Nutrition* 89, no. 6 (June 2009): 1857–63, doi:10.3945/ajcn.2008.26986, epub March 11, 2009.

15. Paul, L., et al., "High Plasma Folate Is Negatively Associated with Leukocyte Telomere Length in Framingham Offspring Cohort," *European Journal of Nutrition* 54, no. 2 (March 2015): 235–41, doi:10.1007 /s00394-014-0704-1.

16. Wojcicki, J., et al., "Early Exclusive Breastfeeding Is Associated with Longer Telomeres in Latino Preschool Children," *American Journal of Clinical Nutrition* (July 20, 2016), doi:10.3945/ajcn.115.115428.

17. Leung, C. W., et al., "Soda and Cell Aging: Associations between Sugar-Sweetened Beverage Consumption and Leukocyte Telomere Length in Healthy Adults from the National Health and Nutrition Examination Surveys," *American Journal of Public Health* 104, no. 12 (December 2014): 2425–31, doi:10.2105/AJPH.2014.302151.

18. Wojcicki, et al "Early Exclusive Breastfeeding Is Associated with Longer Telomeres in Latino Preschool Children." (See #16 above.)

19. "Peppermint Mocha," Starbucks, http://www.starbucks.com/menu/drinks /espresso/peppermint-mocha#size=179560&milk=63&whip=125,accessed September 29, 2015.

20. Pilz, Stefan, Martin Grübler, Martin Gaksch, Verena Schwetz, Christian Trummer, Bríain Ó Hartaigh, Nicolas Verheyen, Andreas Tomaschitz,

and Winfried März. "Vitamin D and Mortality." *Anticancer Research* 36, no. 3 (March 2016): 1379–87.

21. Zhu et al., "Increased Telomerase Activity and Vitamin D Supplementation in Overweight African Americans," *International Journal of Obesity* (June 2012): 805–09, doi:10.1038/ijo.2011.197.

22. Boccardi, V., et al., "Mediterranean Diet, Telomere Maintenance and Health Status Among Elderly," *PLOS ONE* 8, no.4 (April 30, 2013): e62781, doi:10.1371/journal.pone.0062781.

23. Lee, J. Y., et al., "Association Between Dietary Patterns in the Remote Past and Telomere Length," *European Journal of Clinical Nutrition* 69, no. 9 (September 2015): 1048–52, doi:10.1038/ejcn.2015.58.

24. Ibid.

25. "IARC Monographs Evaluate Consumption of Red Meat and Processed Meat," World Health Organization, International Agency for Research on Cancer, press release, October 26, 2015, https://www.iarc.fr/en /media-centre/pr/2015/pdfs/pr240_E.pdf.

26. Nettleton, J. A., et al., "Dietary Patterns, Food Groups, and Telomere Length in the Multi-Ethnic Study of Atherosclerosis (MESA)," *American Journal of Clinical Nutrition* 88, no. 5 (November 2008): 1405–12.

27. Cardin, R., et al., "Effects of Coffee Consumption in Chronic Hepatitis C: A Randomized Controlled Trial," *Digestive and Liver Disease* 45, no. 6 (June 2013): 499–504, doi:10.1016/j.dld.2012.10.021.

28. Liu, J. J., M. Crous-Bou, E. Giovannucci, and I. De Vivo, "Coffee Consumption Is Positively Associated with Longer Leukocyte Telomere Length" in the Nurses' Health Study. *Journal of Nutrition* 146, no. 7 (July 2016): 1373–78, doi:10.3945/jn.116.230490, epub June 8, 2016.

29. Lee, J. Y., et al., "Association Between Dietary Patterns in the Remote Past and Telomere Length" (see #23 above); and Nettleton et al., "Dietary Patterns, Food Groups, and Telomere Length in the Multi-Ethnic Study of Atherosclerosis (MESA)" (see #26 above).

30. García-Calzón, S., et al., "Telomere Length as a Biomarker for Adiposity Changes after a Multidisciplinary Intervention in Overweight/Obese Adolescents: The EVASYON Study," *PLOS ONE* 9, no. 2 (February 24, 2014): e89828, doi:10.1371/journal.pone.0089828.

31. Lee et al., "Association Between Dietary Patterns in the Remote Past and Telomere Length." (See #23 above.)

32. Leung et al., "Soda and Cell Aging." (See #17 above.)

33. Tiainen, A. M., et al., "Leukocyte Telomere Length and Its Relation to Food and Nutrient Intake in an Elderly Population," *European Journal of Clinical Nutrition* 66, no. 12 (December 2012):1290–94, doi:10.1038/ejcn.2012.143.

34. Cassidy, A., et al., "Associations Between Diet, Lifestyle Factors, and Telomere Length in Women," *American Journal of Clinical Nutrition* 91, no. 5 (May 2010): 1273–80, doi:10.3945/ajcn.2009.28947.

35. Pavanello, et al., "Shortened Telomeres in Individuals with Abuse in Alcohol Consumption." (See #3 above.)

36. Cassidy et al., "Associations Between Diet, Lifestyle Factors, and Telomere Length in Women." (See #34 above.)

37. Tiainen et al., "Leukocyte Telomere Length and Its Relation to Food and Nutrient Intake in an Elderly Population." (See #33 above.)

38. Lee et al., "Association Between Dietary Patterns in the Remote Past and Telomere Length." (See #23 above.)

39. Ibid.

40. Ibid.

41. Farzaneh-Far et al., "Association of Marine Omega-3 Fatty Acid Levels With Telomeric Aging in Patients with Coronary Heart Disease." (See #5 above.)

42. García-Calzón et al., "Telomere Length as a Biomarker for Adiposity Changes after a Multidisciplinary Intervention in Overweight/Obese Adolescents: The EVASYON Study." (See #30 above.)

43. Liu et al., "Coffee Consumption Is Positively Associated with Longer Leukocyte Telomere Length" in the Nurses' Health Study. (See #28 above.)

44. Paul, L., "Diet, Nutrition and Telomere Length," *Journal of Nutritional Biochemistry* 22, no. 10 (October 2011): 895–901, doi:10.1016/j.jnutbio.2010.12.001.

45. Richards, J. B., et al., "Higher Serum Vitamin D Concentrations Are Associated with Longer Leukocyte Telomere Length in Women," *American Journal of Clinical Nutrition* 86, no. 5 (November 2007): 1420–25;

46. Xu et al., "Multivitamin Use and Telomere Length in Women" (see #14 above).

47. Paul et al., "High Plasma Folate Is Negatively Associated with Leukocyte Telomere Length in Framingham Offspring Cohort." (This study also found vitamin use was associated with shorter telomeres.) (See #15 above.)

48. O'Neill, J., T. O. Daniel, and L. H. Epstein, "Episodic Future Thinking Reduces Eating in a Food Court," *Eating Behaviors* 20 (January 2016): 9–13, doi:10.1016/j.eatbeh.2015.10.002.

Master Tips for Renewal: Science-Based Suggestions for Making Changes That Last

1. Vasilaki, E. I., S. G. Hosier, and W. M. Cox, "The Efficacy of Motivational Interviewing as a Brief Intervention for Excessive Drinking: A Meta-analytic

Review," *Alcohol and Alcoholism* 41, no. 3 (May 2006): 328–35, doi:10.1093 /alcalc/agl016; and Lindson-Hawley, N., T. P. Thompson, and R. Begh, "Motivational Interviewing for Smoking Cessation," *Cochrane Database of Systematic Reviews* 3 (March 2, 2015): CD006936, doi:10.1002/14651858 .CD006936.pub3.

2. Sheldon, K. M., A. Gunz, C. P. Nichols, and Y. Ferguson, "Extrinsic Value Orientation and Affective Forecasting: Overestimating the Rewards, Under-estimating the Costs," *Journal of Personality* 78, no. 1 (February 2010): 149–78, doi:10.1111/j.1467-6494.2009.00612.x; Kasser, T., and R. M. Ryan, "Further Examining the American Dream: Differential Correlates of Intrinsic and Extrinsic Goals," *Personality and Social Psychology Bulletin* 22, no. 3 (March 1996): 280–87, doi:10.1177/0146167296223006; and Ng, J. Y., et al., "Self-Determination Theory Applied to Health Con-texts: A Meta-analysis," *Perspectives on Psychological Science: A Journal of the Association for Psychological Science* 7, no. 4 (July 2012): 325–40, doi:10.1177/1745691612447309.

3. Ogedegbe, G. O., et al., "A Randomized Controlled Trial of Positive-Affect Intervention and Medication Adherence in Hypertensive African Americans," *Archives of Internal Medicine* 172, no. 4 (February 27, 2012): 322–26, doi:10.1001/archinternmed.2011.1307.

4. Bandura, A., "Self-Efficacy: Toward a Unifying Theory of Behavioral Change." *Psychological Review* 84, no. 2 (March 1977): 191–215.

5. B. J. Fogg illustrates his suggestion of making tiny changes attached to daily trigger events: "Forget Big Change, Start with a Tiny Habit: BJ Fogg at TEDxFremont," YouTube, https://www.youtube.com/watch?v=AdKU Jxjn-R8.

6. Baumeister, R. F., "Self-Regulation, Ego Depletion, and Inhibition," *Neuropsychologia* 65 (December 2014): 313–19, doi:10.1016/j.neuropsycho logia.2014.08.012.

Chapter Eleven: The Places and Faces That Support Our Telomeres

1. Needham, B. L., et al., "Neighborhood Characteristics and Leukocyte Telomere Length: The Multi-ethnic Study of Atherosclerosis," *Health & Place* 28 (July 2014): 167–72, doi:10.1016/j.healthplace.2014.04.009.

2. Geronimus, A. T., et al., "Race-Ethnicity, Poverty, Urban Stress-ors, and Telomere Length in a Detroit Community-Based Sample," *Journal of Health and Social Behavior* 56, no. 2 (June 2015): 199–224, doi:10.1177/0022146515582100.

3. Park, M., et al., "Where You Live May Make You Old: The Associa-tion Between Perceived Poor Neighborhood Quality and Leukocyte

Telomere Length," *PLOS ONE* 10, no. 6 (June 17, 2015): e0128460, doi:10.1371/journal.pone.0128460.

4. Ibid.

5. Lederbogen, F., et al., "City Living and Urban Upbringing Affect Neural Social Stress Processing in Humans," *Nature* 474, no. 7352 (June 22, 2011): 498–501, doi:10.1038/nature10190.

6. Park et al., "Where You Live May Make You Old." (See #3 above.)

7. DeSantis, A. S., et al., "Associations of Neighborhood Characteristics with Sleep Timing and Quality: The Multi-ethnic Study of Atherosclerosis," *Sleep* 36, no. 10 (October 1, 2013): 1543–51, doi:10.5665/sleep.3054.

8. Theall, K. P., et al., "Neighborhood Disorder and Telomeres: Connecting Children's Exposure to Community Level Stress and Cellular Response," *Social Science & Medicine (1982)* 85 (May 2013): 50–58, doi:10.1016/j.socscimed.2013.02.030.

9. Woo, J., et al., "Green Space, Psychological Restoration, and Telomere Length," *Lancet* 373, no. 9660 (January 24, 2009): 299–300, doi:10.1016/S0140-6736(09)60094-5.

10. Roe, J. J., et al., "Green Space and Stress: Evidence from Cortisol Measures in Deprived Urban Communities," *International Journal of Environmental Research and Public Health* 10, no. 9 (September 2013): 4086–103, doi:10.3390/ijerph10094086.

11. Mitchell, R., and F. Popham, "Effect of Exposure to Natural Environment on Health Inequalities: An Observational Population Study," *Lancet* 372, no. 9650 (November 8, 2008): 1655–60, doi:10.1016/S0140-6736(08)61689-X.

12. Theall et al., "Neighborhood Disorder and Telomeres." (See #8 above.)

13. Robertson, T., et al., "Is Socioeconomic Status Associated with Biological Aging as Measured by Telomere Length?" *Epidemiologic Reviews* 35 (2013): 98–111, doi:10.1093/epirev/mxs001.

14. Adler, N. E., et al., "Socioeconomic Status and Health: The Challenge of the Gradient," *American Psychologist* 49, no. 1 (January 1994): 15–24.

15. Cherkas, L. F., et al., "The Effects of Social Status on Biological Aging as Measured by White-Blood-Cell Telomere Length," *Aging Cell* 5, no. 5 (October 2006): 361–65, doi:10.1111/j.1474-9726.2006.00222.x.

16. "Canary Used for Testing for Carbon Monoxide," Center for Construction Research and Training, Electronic Library of Construction Occupational Safety & Health, http://elcosh.org/video/3801/a000096/canary-used-for-testing-for-carbon-monoxide.html.

17. Hou, L., et al., "Lifetime Pesticide Use and Telomere Shortening Among Male Pesticide Applicators in the Agricultural Health Study," *Environ-*

mental Health Perspectives 121, no. 8 (August 2013): 919–24, doi:10.1289 /ehp.1206432.

18. Kahl, V. F., et al., "Telomere Measurement in Individuals Occupation- ally Exposed to Pesticide Mixtures in Tobacco Fields," *Environmental and Molecular Mutagenesis* 57, no. 1 (January 2016), doi:10.1002/em.21984.

19. Ibid.

20. Zota A. R., et al., "Associations of Cadmium and Lead Exposure with Leukocyte Telomere Length: Findings from National Health and Nutri- tion Examination Survey, 1999–2002," *American Journal of Epidemiology* 181, no. 2 (January 15, 2015): 127–136, doi:10.1093/aje/kwu293.

21. "Toxicological Profile for Cadmium," U.S. Department of Health and Human Services, Public Health Service, Agency for Toxic Substances and Disease Registry (Atlanta, Ga., September 2012), http://www.atsdr .cdc.gov/toxprofiles/tp5.pdf.

22. Lin, S., et al., "Short Placental Telomere Was Associated with Cadmium Pollution in an Electronic Waste Recycling Town in China," *PLOS ONE* 8, no. 4 (2013): e60815, doi:10.1371/journal.pone.0060815.

23. Zota et al., "Associations of Cadmium and Lead Exposure with Leuko- cyte Telomere Length." (See #20 above.)

24. Wu, Y., et al., "High Lead Exposure Is Associated with Telomere Length Shortening in Chinese Battery Manufacturing Plant Workers," *Occu- pational and Environmental Medicine* 69, no. 8 (August 2012): 557–63, doi:10.1136/oemed-2011-100478.

25. Ibid.

26. Pawlas, N., et al., "Telomere Length in Children Environmentally Exposed to Low-to-Moderate Levels of Lead," *Toxicology and Applied Pharmacology* 287, no. 2 (September 1, 2015): 111–18, doi:10.1016/j .taap.2015.05.005.

27. Hoxha, M., et al., "Association Between Leukocyte Telomere Shorten- ing and Exposure to Traffic Pollution: A Cross-Sectional Study on Traf- fic Officers and Indoor Office Workers," *Environmental Health* 8 (2009): 41, doi:10.1186/1476-069X-8-41; Zhang, X., S. Lin, W. E. Funk, and L. Hou, "Environmental and Occupational Exposure to Chemicals and Telomere Length in Human Studies," *Postgraduate Medical Jour- nal* 89, no. 1058 (December 2013): 722–28, doi:10.1136/postgradmedj -2012-101350rep; and Mitro, S. D., L. S. Birnbaum, B. L. Needham, and A. R. Zota, "Cross-Sectional Associations Between Exposure to Persis- tent Organic Pollutants and Leukocyte Telomere Length Among U.S. Adults in NHANES, 2001–2002," *Environmental Health Perspectives* 124, no. 5 (May 2016): 651–58, doi:10.1289/ehp.1510187.

28. Bijnens, E., et al., "Lower Placental Telomere Length May Be Attributed to Maternal Residental Traffic Exposure; A Twin Study," *Environment International* 79 (June 2015): 1–7, doi:0.1016/j.envint.2015.02.008.

29. Ferrario, D., et al., "Arsenic Induces Telomerase Expression and Maintains Telomere Length in Human Cord Blood Cells," *Toxicology* 260, nos. 1–3 (June 16, 2009): 132–41, doi:10.1016/j.tox.2009.03.019; Hou, L., et al., "Air Pollution Exposure and Telomere Length in Highly Exposed Subjects in Beijing, China: A Repeated-Measure Study," *Environment International* 48 (November 1, 2012): 71–77, doi:10.1016/j .envint.2012.06.020; Zhang et al., "Environmental and Occupational Exposure to Chemicals and Telomere Length in Human Studies"; Bassig, B. A., et al., "Alterations in Leukocyte Telomere Length in Workers Occupationally Exposed to Benzene," *Environmental and Molecular Mutagenesis* 55, no. 8 (2014): 673–78, doi:10.1002/em.21880; and Li, H., K. Engström, M. Vahter, and K. Broberg, "Arsenic Exposure Through Drinking Water Is Associated with Longer Telomeres in Peripheral Blood," *Chemical Research in Toxicology* 25, no. 11 (November 19, 2012): 2333–39, doi:10.1021/tx300222t.

30. American Association for Cancer Research, *AACR Cancer Progress Report 2014: Transforming Lives Through Cancer Research*, 2014, http://cancer-progressreport.org/2014/Documents/AACR_CPR_2014.pdf, accessed October 21, 2015.

31. "Cancer Fact Sheet No. 297," World Health Organization, updated February 2015,: http://www.who.int/mediacentre/factsheets/fs297/en/, accessed October 21, 2015.

32. House, J. S., K. R. Landis, and D. Umberson, "Social Relationships and Health," *Science* 241, no. 4865 (July 29, 1988): 540–45; Berkman, L. F., and S. L. Syme, "Social Networks, Host Resistance, and Mortality: A Nine-Year Follow-up Study of Alameda County Residents," *American Journal of Epidemiology* 109, no. 2 (February 1979): 186–204; and Holt-Lunstad, J., T. B. Smith, M. B. Baker, T. Harris, and D. Stephenson, "Loneliness and Social Isolation as Risk Factors for Mortality: A Meta-analytic Review," *Perspectives on Psychological Science: A Journal of the Association for Psychological Science* 10, no. 2 (March 2015): 227–37, doi:10.1177/1745691614568352.

33. Hermes, G. L., et al., "Social Isolation Dysregulates Endocrine and Behavioral Stress While Increasing Malignant Burden of Spontaneous Mammary Tumors," *Proceedings of the National Academy of Sciences of the United States of America* 106, no. 52 (December 29, 2009): 22393–98, doi:10.1073/pnas.0910753106.

34. Aydinonat, D., et al., "Social Isolation Shortens Telomeres in African Grey Parrots (*Psittacus erithacus erithacus*)," *PLOS ONE* 9, no. 4 (2014): e93839, doi:10.1371/journal.pone.0093839.

35. Carroll, J. E., A. V. Diez Roux, A. L. Fitzpatrick, and T. Seeman, "Low Social Support Is Associated with Shorter Leukocyte Telomere Length in Late Life: Multi-ethnic Study of Atherosclerosis," *Psychosomatic Medicine* 75, no. 2 (February 2013): 171–77, doi:10.1097/PSY.0b013e31828233bf.

36. Uchino, B. N., et al., "The Strength of Family Ties: Perceptions of Network Relationship Quality and Levels of C-Reactive Proteins in the North Texas Heart Study," *Annals of Behavioral Medicine* 49, no. 5 (October 2015): 776–81, doi:10.1007/s12160-015-9699-y.

37. Uchino, B. N., et al., "Social Relationships and Health: Is Feeling Positive, Negative, or Both (Ambivalent) About Your Social Ties Related to Telomeres?" *Health Psychology* 31, no. 6 (November 2012): 789–96, doi:10.1037/a0026836.

38. Robles, T. F., R. B. Slatcher, J. M. Trombello, and M. M. McGinn, "Marital Quality and Health: A Meta-analytic Review," *Psychological Bulletin* 140, no. 1 (January 2014): 140–87, doi:10.1037/a0031859.

39. Ibid.

40. Mainous, A. G., et al., "Leukocyte Telomere Length and Marital Status among Middle-Aged Adults," *Age and Ageing* 40, no. 1 (January 2011): 73–78, doi:10.1093/ageing/afq118; and Yen, Y., and F. Lung, "Older Adults with Higher Income or Marriage Have Longer Telomeres," *Age and Ageing* 42, no. 2 (March 2013): 234–39, doi:10.1093/ageing/afs122.

41. Broer, L., V. Codd, D. R. Nyholt, et al, "Meta-Analysis of Telomere Length in 19,713 Subjects Reveals High Heritability, Stronger Maternal Inheritance and a Paternal Age Effect," *European Journal of Human Genetics: EJHG* 21, no. 10 (October 2013): 1163–68, doi:10.1038/ejhg.2012.303.

42. Herbenick, D., et al., "Sexual Behavior in the United States: Results from a National Probability Sample of Men and Women Ages 14–94," *Journal of Sexual Medicine* 7, Suppl. 5 (October 7, 2010): 255–65, doi:10.1111/j.1743-6109.2010.02012.x.

43. Saxbe, D. E., et al., "Cortisol Covariation within Parents of Young Children: Moderation by Relationship Aggression," *Psychoneuroendocrinology* 62 (December 2015): 121–28, doi:10.1016/j.psyneuen.2015.08.006.

44. Liu, S., M. J. Rovine, L. C. Klein, and D. M. Almeida, "Synchrony of Diurnal Cortisol Pattern in Couples," *Journal of Family Psychology* 27, no. 4 (August 2013): 579–88, doi:10.1037/a0033735.

45. Helm, J. L., D. A. Sbarra, and E. Ferrer, "Coregulation of Respiratory

Sinus Arrhythmia in Adult Romantic Partners," *Emotion* 14, no. 3 (June 2014): 522–31, doi:10.1037/a0035960.

46. Hack, T., S. A. Goodwin, and S. T. Fiske, "Warmth Trumps Competence in Evaluations of Both Ingroup and Outgroup," *International Journal of Science, Commerce and Humanities* 1, no. 6 (September 2013): 99–105.

47. Parrish, T., "How Hate Took Hold of Me," *Daily News*, June 21, 2015, http://www.nydailynews.com/opinion/tim-parrish-hate-hold-article-1.2264643, accessed October 23, 2015.

48. Lui, S. Y., and Kawachi, I. "Discrimination and Telomere Length Among Older Adults in the US: Does the Association Vary by Race and Type of Discrimination?" under review, Public Health Reports.

49. Chae, D. H., et al., "Discrimination, Racial Bias, and Telomere Length in African American Men," *American Journal of Preventive Medicine* 46, no. 2 (February 2014): 103–11, doi:10.1016/j.amepre.2013.10.020.

50. Peckham, M., "This Billboard Sucks Pollution from the Sky and Returns Purified Air," *Time*, May 1, 2014, http://time.com/84013/this-billboard-sucks-pollution-from-the-sky-and-returns-purified-air/, accessed November 24, 2015.

51. Diers, J., *Neighbor Power: Building Community the Seattle Way* (Seattle: University of Washington Press, 2004).

52. Beyer, K. M. M., et al., "Exposure to Neighborhood Green Space and Mental Health: Evidence from the Survey of the Health of Wisconsin," *International Journal of Environmental Research and Public Health* 11, no. 3 (March 2014): 3453–72, doi:10.3390/ijerph110303453; and Roe et al., "Green Space and Stress" (see #10 above).

53. Branas, C. C., et al., "A Difference-in-Differences Analysis of Health, Safety, and Greening Vacant Urban Space," *American Journal of Epidemiology* 174, no. 11 (December 1, 2011): 1296–1306, doi:10.1093/aje/kwr273.

54. Wesselmann, E. D., F. D. Cardoso, S. Slater, and K. D. Williams, "To Be Looked At as Though Air: Civil Attention Matters," *Psychological Science* 23, no. 2 (February 2012): 166–168, doi:10.1177/0956797611427921.

55. Guéguen, N., and M-A De Gail, "The Effect of Smiling on Helping Behavior: Smiling and Good Samaritan Behavior," *Communication Reports*, 16, no. 2 (2003): 133–40, doi: 10.1080/08934210309384496.

Chapter Twelve: Pregnancy: Cellular Aging Begins in the Womb

1. Hjelmborg, J. B., et al., "The Heritability of Leucocyte Telomere Length Dynamics," *Journal of Medical Genetics* 52, no. 5 (May 2015): 297–302, doi:10.1136/jmedgenet-2014-102736.

2. Wojcicki, J. M., et al., "Cord Blood Telomere Length in Latino Infants: Relation with Maternal Education and Infant Sex," *Journal of Perinatology: Official Journal of the California Perinatal Association* 36, no. 3 (March 2016): 235–41, doi:10.1038/jp.2015.178.

3. Needham, B. L., et al., "Socioeconomic Status and Cell Aging in Children," *Social Science and Medicine (1982)* 74, no. 12 (June 2012): 1948–51, doi:10.1016/j.socscimed.2012.02.019.

4. Collopy, L. C., et al., "Triallelic and Epigenetic-like Inheritance in Human Disorders of Telomerase," *Blood* 126, no. 2 (July 9, 2015): 176–84, doi:10.1182/blood-2015-03-633388.

5. Factor-Litvak, P., et al., "Leukocyte Telomere Length in Newborns: Implications for the Role of Telomeres in Human Disease," *Pediatrics* 137, no. 4 (April 2016): e20153927, doi:10.1542/peds.2015-3927.

6. De Meyer, T., et al., "A Non-Genetic, Epigenetic-like Mechanism of Telomere Length Inheritance?" *European Journal of Human Genetics* 22, no. 1 (January 2014): 10–11, doi:10.1038/ejhg.2013.255.

7. Collopy et al., "Triallelic and Epigenetic-like Inheritance in Human Disorders of Telomerase." (See #4 above.)

8. Tarry-Adkins, J. L., et al., "Maternal Diet Influences DNA Damage, Aortic Telomere Length, Oxidative Stress, and Antioxidant Defense Capacity in Rats," *FASEB Journal: Official Publication of the Federation of American Societies for Experimental Biology* 22, no. 6 (June 2008): 2037–44, doi:10.1096/fj.07-099523.

9. Aiken, C. E., J. L. Tarry-Adkins, and S. E. Ozanne, "Suboptimal Nutrition in Utero Causes DNA Damage and Accelerated Aging of the Female Reproductive Tract," *FASEB Journal: Official Publication of the Federation of American Societies for Experimental Biology* 27, no. 10 (October 2013): 3959–65, doi:10.1096/fj.13-234484.

10. Aiken, C. E., J. L. Tarry-Adkins, and S. E. Ozanne. "Transgenerational Developmental Programming of Ovarian Reserve," *Scientific Reports* 5 (2015): 16175, doi:10.1038/srep16175.

11. Tarry-Adkins, J. L., et al., "Nutritional Programming of Coenzyme Q: Potential for Prevention and Intervention?" *FASEB Journal: Official Publication of the Federation of American Societies for Experimental Biology* 28, no. 12 (December 2014): 5398–405, doi:10.1096/fj.14-259473.

12. Bull, C., H. Christensen, and M. Fenech, "Cortisol Is Not Associated with Telomere Shortening or Chromosomal Instability in Human Lymphocytes Cultured Under Low and High Folate Conditions," *PLOS ONE* 10, no. 3 (March 6, 2015): e0119367, doi:10.1371/journal.pone.0119367; and Bull, C., et al., "Folate Deficiency Induces Dysfunctional Long and Short Telomeres;

Both States Are Associated with Hypomethylation and DNA Damage in Human WIL2-NS Cells," *Cancer Prevention Research (Philadelphia, Pa.)* 7, no. 1 (January 2014): 128–38, doi:10.1158/1940-6207.CAPR-13-0264.

13. Entringer, S., et al., "Maternal Folate Concentration in Early Pregnancy and Newborn Telomere Length," *Annals of Nutrition and Metabolism* 66, no. 4 (2015): 202–08, doi:10.1159/000381925.

14. Cerne, J. Z., et al., "Functional Variants in CYP1B1, KRAS and MTHFR Genes Are Associated with Shorter Telomere Length in Post-menopausal Women," *Mechanisms of Ageing and Development* 149 (July 2015): 1–7, doi:10.1016/j.mad.2015.05.003.

15. "Folic Acid Fact Sheet," Womenshealth.gov, http://womenshealth.gov /publications/our-publications/fact-sheet/folic-acid.html, accessed November 27, 2015.

16. Paul, L., et al., "High Plasma Folate Is Negatively Associated with Leukocyte Telomere Length in Framingham Offspring Cohort," *European Journal of Nutrition* 54, no. 2 (March 2015): 235–41, doi:10.1007/s00394-014-0704-1.

17. Entringer, S., et al., "Maternal Psychosocial Stress During Pregnancy Is Associated with Newborn Leukocyte Telomere Length," *American Journal of Obstetrics and Gynecology* 208, no. 2 (February 2013): 134.e1–7, doi:10.1016/j.ajog.2012.11.033.

18. Marchetto, N. M., et al., "Prenatal Stress and Newborn Telomere Length," *American Journal of Obstetrics and Gynecology*, January 30, 2016, doi:10.1016/j.ajog.2016.01.177.

19. Entringer, S., et al., "Influence of Prenatal Psychosocial Stress on Cytokine Production in Adult Women," *Developmental Psychobiology* 50, no. 6 (September 2008): 579–87, doi:10.1002/dev.20316.

20. Entringer, S., et al., "Stress Exposure in Intrauterine Life Is Associated with Shorter Telomere Length in Young Adulthood," *Proceedings of the National Academy of Sciences of the United States of America* 108, no. 33 (August 16, 2011): E513–18, doi:10.1073/pnas.1107759108.

21. Haussman, M., and B. Heidinger, "Telomere Dynamics May Link Stress Exposure and Ageing across Generations," *Biology Letters* 11, no. 11 (November 2015), doi:10.1098/rsbl.2015.0396.

22. Ibid.

Chapter Thirteen: Childhood Matters for Life: How the Early Years Shape Telomeres

1. Sullivan, M. C.," For Romania's Orphans, Adoption Is Still a Rarity," National Public Radio, August 19, 2012, http://www.npr.org/2012/08/19 /158924764/for-romanias-orphans-adoption-is-still-a-rarity.

2. Ahern, L., "Orphanages Are No Place for Children," *Washington Post*, August 9, 2013, https://www.washingtonpost.com/opinions/orphanages -are-no-place-for-children/2013/08/09/6d502fb0-fadd-11e2-a369-d1954 abcb7e3_story.html, accessed October 14, 2015.

3. Felitti, V. J., et al., "Relationship of Childhood Abuse and Household Dysfunction to Many of the Leading Causes of Death in Adults: The Adverse Childhood Experiences (ACE) Study," *American Journal of Preventive Medicine* 14, no. 4 (May 1998): 245–58.

4. Chen, S. H., et al., "Adverse Childhood Experiences and Leukocyte Telomere Maintenance in Depressed and Healthy Adults," *Journal of Affective Disorders* 169 (December 2014): 86–90, doi:10.1016/j.jad.2014 .07.035.

5. Skilton, M. R., et al., "Telomere Length in Early Childhood: Early Life Risk Factors and Association with Carotid Intima-Media Thickness in Later Childhood," *European Journal of Preventive Cardiology* 23, no. 10 (July 2016), 1086–92, doi:10.1177/2047487315607075.

6. Drury, S. S., et al., "Telomere Length and Early Severe Social Deprivation: Linking Early Adversity and Cellular Aging," *Molecular Psychiatry* 17, no. 7 (July 2012): 719–27, doi:10.1038/mp.2011.53.

7. Hamilton, J., "Orphans' Lonely Beginnings Reveal How Parents Shape a Child's Brain," National Public Radio, February 24, 2014, http://www .npr.org/sections/health-shots/2014/02/20/280237833/orphans -lonely-beginnings-reveal-how-parents-shape-a-childs-brain, accessed October 15, 2015.

8. Powell, A., "Breathtakingly Awful," *Harvard Gazette*, October 5, 2010, http://news.harvard.edu/gazette/story/2010/10/breathtakingly-awful/, accessed October 26, 2015.

9. Authors' interview with Charles Nelson, September 18, 2015.

10. Shalev, I., et al., "Exposure to Violence During Childhood Is Associated with Telomere Erosion from 5 to 10 Years of Age: A Longitudinal Study," *Molecular Psychiatry* 18, no. 5 (May 2013): 576–81, doi:10.1038 /mp.2012.32.

11. Price, L. H., et al., "Telomeres and Early-Life Stress: An Overview," *Biological Psychiatry* 73, no. 1 (January 1, 2013): 15–23, doi:10.1016/j .biopsych.2012.06.025.

12. Révész, D., Y. Milaneschi, E. M. Terpstra, and B. W. J. H. Penninx, "Baseline Biopsychosocial Determinants of Telomere Length and 6-Year Attrition Rate," *Psychoneuroendocrinology* 67 (May 2016): 153–62, doi:10 .1016/j.psyneuen.2016.02.007.

13. Danese, A., and B. S. McEwen, "Adverse Childhood Experiences,

Allostasis, Allostatic Load, and Age-Related Disease," *Physiology & Behavior* 106, no. 1 (April 12, 2012): 29–39, doi:10.1016/j.physbeh.2011.08.019.

14. Infurna, F. J., C. T. Rivers, J. Reich, and A. J. Zautra, "Childhood Trauma and Personal Mastery: Their Influence on Emotional Reactivity to Everyday Events in a Community Sample of Middle-Aged Adults," *PLOS ONE* 10, no. 4 (2015): e0121840, doi:10.1371/journal.pone.0121840.

15. Schrepf, A., K. Markon, and S. K. Lutgendorf, "From Childhood Trauma to Elevated C-Reactive Protein in Adulthood: The Role of Anxiety and Emotional Eating," *Psychosomatic Medicine* 76, no. 5 (June 2014): 327–36, doi:10.1097/PSY.0000000000000072.

16. Felitti, V. J., et al., "Relationship of Childhood Abuse and Household Dysfunction to Many of the Leading Causes of Death in Adults. The Adverse Childhood Experiences (ACE) Study," *American Journal of Preventive Medicine* 14, no. 4 (May 1998): 245–58, doi.org/10.1016/S0749-3797(98)00017-8.

17. Lim, D., and D. DeSteno, "Suffering and Compassion: The Links Among Adverse Life Experiences, Empathy, Compassion, and Prosoial Behavior," *Emotion* 16, no. 2 (March 2016): 175–82, doi:10.1037/emo0000144.

18. Asok, A., et al., "Infant-Caregiver Experiences Alter Telomere Length in the Brain," *PLOS ONE* 9, no. 7 (2014): e101437, doi:10.1371/journal.pone.0101437.

19. McEwen, B. S., C. N. Nasca, and J. D. Gray, "Stress Effects on Neuronal Structure: Hippocampus, Amygdala, and Prefrontal Cortex," *Neuropsychopharmacology: Official Publication of the American College of Neuropsychopharmacology* 41, no. 1 (January 2016): 3–23, doi:10.1038/npp.2015.171; and Arnsten, A. F. T., "Stress Signalling Pathways That Impair Prefrontal Cortex Structure and Function," *Nature Reviews Neuroscience* 10, no. 6 (June 2009): 410–22, doi:10.1038/nrn2648.

20. Suomi, S., "Attachment in Rhesus Monkeys," in *Handbook of Attachment: Theory, Research, and Clinical Applications*, ed. J. Cassidy and P. R. Shaver, 3rd ed. (New York: Guilford Press, 2016).

21. Schneper, L., Jeanne Brooks-Gunn, Daniel Notterman, and Stephen, Suomi, "Early Life Experiences and Telomere Length in Adult Rhesus Monkeys: An Exploratory Study." *Psychosomatic Medicine*, in press (n.d.).

22. Gunnar, M. R., et al., "Parental Buffering of Fear and Stress Neurobiology: Reviewing Parallels Across Rodent, Monkey, and Human Models," *Social Neuroscience* 10, no. 5 (2015): 474–78, doi:10.1080/17470919.2015.1070198.

23. Hostinar, C. E., R. M. Sullivan, and M. R. Gunnar, "Psychobiological Mechanisms Underlying the Social Buffering of the Hypothalamic-Pituitary-Adrenocortical Axis: A Review of Animal Models and Human

Studies Across Development," *Psychological Bulletin* 140, no. 1 (January 2014): 256–82, doi:10.1037/a0032671.

24. Doom, J. R., C. E. Hostinar, A. A. VanZomeren-Dohm, and M. R. Gunnar, "The Roles of Puberty and Age in Explaining the Diminished Effectiveness of Parental Buffering of HPA Reactivity and Recovery in Adolescence," *Psychoneuroendocrinology* 59 (September 2015): 102–11, doi:10.1016/j.psyneuen.2015.04.024.

25. Seery, M. D., et al., "An Upside to Adversity?: Moderate Cumulative Lifetime Adversity Is Associated with Resilient Responses in the Face of Controlled Stressors," *Psychological Science* 24, no. 7 (July 1, 2013): 1181–89, doi:10.1177/0956797612469210.

26. Asok, A., et al., "Parental Responsiveness Moderates the Association Between Early-Life Stress and Reduced Telomere Length," *Development and Psychopathology* 25, no. 3 (August 2013): 577–85, doi:10.1017/S0954579413000011.

27. Bernard, K., C. E. Hostinar, and M. Dozier, "Intervention Effects on Diurnal Cortisol Rhythms of Child Protective Services–Referred Infants in Early Childhood: Preschool Follow-Up Results of a Randomized Clinical Trial," *JAMA Pediatrics* 169, no. 2 (February 2015): 112–19, doi:10.1001/jamapediatrics.2014.2369.

28. Kroenke, C. H., et al., "Autonomic and Adrenocortical Reactivity and Buccal Cell Telomere Length in Kindergarten Children," *Psychosomatic Medicine* 73, no. 7 (September 2011): 533–40, doi:10.1097/PSY.0b013e318229acfc.

29. Wojcicki, J. M., et al., "Telomere Length Is Associated with Oppositional Defiant Behavior and Maternal Clinical Depression in Latino Preschool Children," *Translational Psychiatry* 5 (June 2015): e581, doi:10.1038/tp.2015.71; and Costa, D. S., et al., "Telomere Length Is Highly Inherited and Associated with Hyperactivity-Impulsivity in Children with Attention Deficit/Hyperactivity Disorder," *Frontiers in Molecular Neuroscience* 8 (July 2015): 28, doi:10.3389/fnmol.2015.00028.

30. Kroenke et al., "Autonomic and Adrenocortical Reactivity and Buccal Cell Telomere Length in Kindergarten Children." (See #27 above.)

31. Boyce, W. T., and B. J. Ellis, "Biological Sensitivity to Context: I. An Evolutionary-Developmental Theory of the Origins and Functions of Stress Reactivity," *Development and Psychopathology* 17, no. 2 (spring 2005): 271–301.

32. Van Ijzendoorn, M. H., and M. J. Bakermans-Kranenburg, "Genetic Differential Susceptibility on Trial: Meta-analytic Support from Randomized Controlled Experiments," *Development and Psychopathology* 27, no. 1 (February 2015): 151–62, doi:10.1017/S0954579414001369.

33. Colter, M., et al., "Social Disadvantage, Genetic Sensitivity, and Children's Telomere Length," *Proceedings of the National Academy of Sciences of the United States of America* 111, no. 16 (April 22, 2014): 5944–49, doi:10.1073/pnas.1404293111.

34. Brody, G. H., T. Yu, S. R. H. Beach, and R. A. Philibert, "Prevention Effects Ameliorate the Prospective Association Between Nonsupportive Parenting and Diminished Telomere Length," *Prevention Science: The Official Journal of the Society for Prevention Research* 16, no. 2 (February 2015): 171–80, doi:10.1007/s11121-014-0474-2; Beach, S. R. H., et al., "Nonsupportive Parenting Affects Telomere Length in Young Adulthood Among African Americans: Mediation through Substance Use," *Journal of Family Psychology: JFP: Journal of the Division of Family Psychology of the American Psychological Association (Division 43)* 28, no. 6 (December 2014): 967–72, doi:10.1037/fam0000039; and Brody, G. H., et al., "The Adults in the Making Program: Long-Term Protective Stabilizing Effects on Alcohol Use and Substance Use Problems for Rural African American Emerging Adults," *Journal of Consulting and Clinical Psychology* 80, no. 1 (February 2012): 17–28. doi:10.1037/a0026592.

35. Brody et al., "Prevention Effects Ameliorate the Prospective Association Between Nonsupportive Parenting and Diminished Telomere Length"; and Beach et al., "Nonsupportive Parenting Affects Telomere Length in Young Adulthood among African Americans: Mediation through Substance Use." (See #33 above.)

36. Spielberg, J. M., T. M. Olino, E. E. Forbes, and R. E. Dahl, "Exciting Fear in Adolescence: Does Pubertal Development Alter Threat Processing?" *Developmental Cognitive Neuroscience* 8 (April 2014): 86–95, doi:10.1016/j.dcn.2014.01.004; and Peper, J. S., and R. E. Dahl, "Surging Hormones: Brain-Behavior Interactions During Puberty," *Current Directions in Psychological Science* 22, no. 2 (April 2013): 134–39, doi:10.1177/0963721412473755.

37. Turkle, S., *Reclaiming Conversation: The Power of Talk in a Digital Age* (New York: Penguin Press, 2015).

38. Siegel, D., and T. P. Bryson, *The Whole-Brain Child: 12 Revolutionary Strategies to Nurture Your Child's Developing Mind* (New York: Delacorte Press, 2011).

39. Robles, T. F., et al., "Emotions and Family Interactions in Childhood: Associations with Leukocyte Telomere Length Emotions, Family Interactions, and Telomere Length," *Psychoneuroendocrinology* 63 (January 2016): 343–50, doi:10.1016/j.psyneuen.2015.10.018.

Conclusion: Entwined: Our Cellular Legacy

1. Pickett, K. E., and R. G. Wilkinson, "Inequality: An Underacknowledged Source of Mental Illness and Distress," *British Journal of Psychiatry: The Journal of Mental Science* 197, no. 6 (December 2010): 426–28, doi:10.1192/bjp.bp.109.072066.

2. Ibid; and Wilkerson, R. G., and K. Pickett, *The Spirit Level: Why More Equal Societies Almost Always Do Better* (London: Allen Lane, 2009).

3. Stone, C., D. Trisi, A. Sherman, and B. Debot, "A Guide to Statistics on Historical Trends in Income Inequality," Center on Budget and Policy Priorities, updated October 26, 2015, http://www.cbpp.org/research/poverty-and-inequality/a-guide-to-statistics-on-historical-trends-in-income-inequality.

4. Pickett, K. E., and R. G. Wilkinson, "The Ethical and Policy Implications of Research on Income Inequality and Child Wellbeing," *Pediatrics* 135, Suppl. 2 (March 2015): S39–47, doi:10.1542/peds.2014-3549E.

5. Mayer, E. A., et al., "Gut Microbes and the Brain: Paradigm Shift in Neuroscience," *Journal of Neuroscience: The Official Journal of the Society for Neuroscience* 34, no. 46 (November 12, 2014): 15490–96, doi:10.1523/JNEUROSCI.3299-14.2014; Picard, M., R. P. Juster, and B. S. McEwen, "Mitochondrial Allostatic Load Puts the 'Gluc' Back in Glucocorticoids," *Nature Reviews Endocrinology* 10, no. 5 (May 2014): 303–10, doi:10.1038/nrendo.2014.22; and Picard, M., et al., "Chronic Stress and Mitochondria Function in Humans," under review.

6. Varela, F. J., E. Thompson, and E. Rosch, *The Embodied Mind* (Cambridge, MA: MIT Press, 1991).

7. "Zuckerberg: One in Seven People on the Planet Used Facebook on Monday," *Guardian*, August 28, 2015, http://www.theguardian.com/technology/2015/aug/27/facebook-1bn-users-day-mark-zuckerberg, accessed October 26, 2015; and "Number of Monthly Active Facebook Users Worldwide as of 1st Quarter 2016 (in Millions)," Statista, http://www.statista.com/statistics/264810/number-of-monthly-active-facebook-users-worldwide/.

We thank the many authors and organizations that allowed us permissions to reprint scales and figures.

For figures, this includes:

Blackburn, Elizabeth H., Elissa S. Epel, and Jue Lin. "Human Telomere Biology: A Contributory and Interactive Factor in Aging, Disease Risks, and Protection." *Science* (New York, N.Y.) 350, no. 6265 (December 4, 2015): 1193–98. **Reprinted with permission from AAAS.**

Epel, Elissa S., Elizabeth H. Blackburn, Jue Lin, Firdaus S. Dhabhar, Nancy E. Adler, Jason D. Morrow, and Richard M. Cawthon. "Accelerated Telomere Shortening in Response to Life Stress." *Proceedings of the National Academy of Sciences of the United States of America* 101, no. 49 (December 7, 2004): 17312–15. **Permissions granted by the National Academy of Sciences, U.S.A. Copyright (2004) National Academy of Sciences, U.S.A.**

Cribbet, M. R., M. Carlisle, R. M. Cawthon, B. N. Uchino, P. G. Williams, T. W. Smith, and K. C. Light. "Cellular Aging and Restorative Processes: Subjective Sleep Quality and Duration Moderate the Association between Age and Telomere Length in a Sample of Middle-Aged and Older Adults." *SLEEP* 37, no. 1: 65–70. **Republished with permission of the American Academy of Sleep Medicine; permission conveyed through Copyright Clearance Center, Inc.**

Carroll J. E., S. Esquivel, A. Goldberg, T. E. Seeman, R. B. Effros, J. Dock, R. Olmstead, E. C. Breen, and M. R. Irwin. "Insomnia and Telomere Length in Older Adults." *SLEEP* 39, no 3 (2016): 559–64. **Republished with permission of the American Academy of Sleep Medicine; permission conveyed through Copyright Clearance Center, Inc.**

Farzaneh-Far R, J. Lin, E. S. Epel, W. S. Harris, E. H. Blackburn, and M. A. Whooley. "Association of Marine Omega-3 Fatty Acid Levels with Telomeric Aging in Patients with Coronary Heart Disease." *JAMA* 303, no 3 (2010): 250–57. **Permissions granted by the American Medical Association.**

Park, M., J. E. Verhoeven, P. Cuijpers, C. F. Reynolds III, and B. W. J. H. Penninx. "Where You Live May Make You Old: The Association between Perceived Poor Neighborhood Quality and Leukocyte Telomere Length." *PLoS ONE* 10, no.6 (2015), e0128460. http://doi. org/10.1371/journal.pone.0128460. **Permissions granted by Park et al. via the Creative Commons Attribution License. Copyright © 2015 Park et al.**

Brody, G. H., T. Yu, S. R. H. Beach, and R. A. Philibert. "Prevention Effects Ameliorate the Prospective Association between Nonsupportive Parenting and Diminished Telomere Length." *Prevention Science: The Official*

Journal of the Society for Prevention Research 16, no. 2 (February 2015): 171–80. **With permission of Springer.**

Pickett, Kate E., and Richard G. Wilkinson. "Inequality: An Underacknowledged Source of Mental Illness and Distress." *The British Journal of Psychiatry: The Journal of Mental Science* 197, no. 6 (December 2010): 426–28. **Permissions granted by the Royal College of Psychiatrists. Copyright, the Royal College of Psychiatrists.**

For scales, this includes:

Kiernan, M., D. E. Schoffman, K. Lee, S. D. Brown, J. M. Fair, M. G. Perri, and W. L. Haskell. "The Stanford Leisure-Time Activity Categorical Item (L-Cat): A Single Categorical Item Sensitive to Physical Activity Changes in Overweight/Obese Women." *International Journal of Obesity* 37 (2013): 1597–602. **Permissions granted by Nature Publishing Group and Dr. Michaela Kiernan, Stanford University School of Medicine. Copyright 2013. Reprinted by permission from Macmillan Publishers Ltd.**

The ENRICHD Investigators. "Enhancing Recovery in Coronary Heart Disease (ENRICHD): Baseline Characteristics." *The American Journal of Cardiology* 88, no. 3, (August 1, 2001): 316–22. **Permissions granted by Elsevier science and technology journals and Dr. Pamela Mitchell, University of Washington. Permission conveyed through Copyright Clearance Center, Inc. Republished with permission of Elsevier Science and Technology Journals.**

Buysse, Daniel J., Charles F. Reynolds III, Timothy H. Monk, Susan R. Berman, and David J. Kupfer. "The Pittsburgh Sleep Quality Index: A New Instrument for Psychiatric Practice and Research." *Psychiatry Research* 28, no. 2 (May 1989): 193–213. **Copyright © 1989 and 2010, University of Pittsburgh. All rights reserved. Permissions granted by Dr. Daniel Buysse and the University of Pittsburgh.**

Scheier, M. F., and C. S. Carver. "Optimism, Coping, and Health: Assessment and Implications of Generalized Outcome Expectancies." *Health Psychology* 4, no. 3 (1985): 219–47. **Permissions granted by Dr. Michael Scheier, Carnegie Mellon University, and the American Psychological Association.**

Trapnell, P. D., J. D. Campbell. "Private Self-Consciousness and the Five-Factor Model of Personality: Distinguishing Rumination from Reflection." *Journal of Personality and Social Psychology* 76 (1999): 284–330. **Permissions granted by Dr. Paul Trapnell, University of Winnipeg, and the American Psychological Association.**

John, O. P., E. M. Donahue, and R. L. Kentle. Conscientiousness: "The Big Five Inventory—Versions 4a and 54." Berkeley: University of California, Berkeley, Institute of Personality and Social Research, 1991. **Permissions granted by Dr. Oliver John, University of California, Berkeley.**

Scheier, M. F., C. Wrosch, A. Baum, S. Cohen, L. M. Martire, K. A. Matthews, R. Schulz, and B. Zdaniuk. "The Life Engagement Test: Assessing Purpose in Life." *Journal of Behavioral Medicine* 29 (2006): 291–98. **With permission of Springer. Permissions granted by Springer Publishing and Dr. Michael Scheier, Carnegie Mellon University.**

The Adverse Childhood Experiences Scale (ACES) was reprinted with permission from Dr. Vincent Felitti, MD, Co-PI, Adverse Childhood Experiences Study, University of California, San Diego.

Index

Page numbers in italics refer to illustrations, charts, and graphs in the text.

About the Authors

Dr. Elizabeth Blackburn, PhD, received the Nobel Prize in Physiology or Medicine in 2009 alongside two colleagues for the discovery of the molecular nature of telomeres, the ends of chromosomes that serve as protective caps, and for discovering telomerase, the enzyme that maintains telomeres. She is currently the president of the Salk Institute and professor emeritus at UCSF. Blackburn is a past president of the American Association for Cancer Research and the American Society for Cell Biology and is a recipient of nearly every major medical award, including the Albert Lasker Basic Medical Research Award. She was named one of *TIME* magazine's 100 most influential people. She is a member of the U.S. National Academies of Sciences and Medicine and the Royal Society of London. Blackburn has helped guide public science policy and served on the President's Council on Bioethics, an advisory committee to the president of the United States.

Blackburn was born in Tasmania, Australia. She received her bachelor of science degree from the University of Melbourne and her PhD in molecular biology from the University of Cambridge and conducted her postdoctoral fellowship at Yale University. She and her husband currently live in La Jolla, California, and part-time in San Francisco.

Dr. Elissa Epel, PhD, is a leading health psychologist who studies stress, aging, and obesity. She is a professor in the Department of Psychiatry at UCSF, the director of UCSF's Aging, Metabolism, and Emotions (AME) Center, director of COAST, a UCSF obesity research center, and associate director of UCSF's Center for Health and Community. She is a member of the National Academy of Medicine and serves on scientific advisory committees for National Institute of Health initiatives (such as the Science of Behavior Change program), the Mind & Life Institute, and the European Society of Preventive Medicine. She has received many research awards, including awards from Stanford University, the Society of Behavioral Medicine, the Academy of Behavioral Medicine Research, and the American Psychological Association.

Epel was born in Carmel, California. She received her bachelor's degree from Stanford University and her PhD in clinical and health psychology from Yale University. She completed her clinical internship at the Veterans Administration Palo Alto Healthcare System and conducted her postdoctoral fellowship at UCSF. She lives in San Francisco with her husband and son.